EVERYTHING YOU NE

AMATEUR RADIO SATELLITES
FOR BEGINNERS

By Steve Ford, WB8IMY
Jodi Morin, KA1JPA
Michelle Bloom, WB1ENT
David Pingree, N1NAS
Maty Weinberg, KB1EIB

ARRL
The national association for
AMATEUR RADIO®

Copyright © 2020 by

The American Radio Relay League

*Copyright secured under the
Pan-American Convention*

International Copyright secured

All rights reserved. No part of this work may be reproduced in any form except with written permission of the publisher. All rights of translation are reserved.

Printed in USA

Quedan reservados todos los derechos

ISBN: 978-1-62595-130-4

First Edition
First Printing

On our cover:

Adam Whitney, KØFFY, made contacts through the AMSAT-OSCAR 92 satellite from the rim of the caldera of the Hverfjall volcano in northern Iceland.
[Jessica Whitney, photo]

Table of Contents

Chapter 1
The Rich History of Amateur Satellites
Hams have been at this for more than 60 years! See how far we've come.

Chapter 2
Mysteries of Where and When
Not only do you need to know where the satellite is, you need to know where *you* are.

Chapter 3
Your Satellite Station
It can be simple or complex. Read this helpful guide for making the best choices.

Chapter 4
Let's Get on the Air!
It's time to turn on your radio and start making contacts. Find out how!

Chapter 5
Satellites in Education
Put your satellite skills to work promoting amateur radio and supporting education.

Chapter 6
Antenna System Projects
Save money by building your own antennas, along with an inexpensive az/el rotator to move them.

Foreword

Building amateur radio satellites is difficult; communicating through amateur radio satellites is not. But what makes communication through satellites so different from other sections of our great hobby? The challenge (and enjoying it)!

The challenge involves expanding your horizons to include physics, mathematics, orbital mechanics, and a deeper understanding of the radio science. As mentors, ARRL is providing this book as your starting point, and is a partner with AMSAT, the Radio Amateur Satellite Corporation, in its mission of Keeping Amateur Radio in Space™ by building and launching amateur radio satellites.

The year 2019 was the 50th anniversary of AMSAT. The pioneers of Amateur Radio in Space™ began their work more than 60 years ago with Project OSCAR, our first satellite. While there have been many firsts for amateur radio along the way, two stand out dramatically.

The first to mention is AMSAT-OSCAR-7. This satellite has been in space since November 15, 1974. At the time, the predictions from "experts" gave the satellite about 3 weeks of life before it would be destroyed by radiation. The fact that OSCAR 7 is still operational is a testimony to the skills of the hams that built her. Today, it is the oldest operational satellite *ever*!

The second is the SmallSat Revolution. Satellites don't have to be large, hulking spacecraft and AMSAT proved this with launches beginning in the early '90s from Space Launch Complex 2 West (SLC-2W) at Vandenberg Air Force Base in California. From here we sent compact, innovative satellites into orbit. Commercial and military satellite designers watched our success and soon followed.

AMSAT is pleased that ARRL is publishing this book for the amateur community. Our partnership has been long and fruitful, and we support ARRL's work to champion amateur radio interests with the International Amateur Radio Union, the International Telecommunications Union, and the Federal Communications Commission. Amateur Radio on the International Space Station (ARISS) is also an AMSAT/ARRL partnership.

Your path to space communications can take many forms, including learning satellite basics, how to locate satellites, what antenna and radio systems are available and work best for satellite uses, the operation of FM, SSB/CW, or linear satellites, what digital modes (JT-65, BPSK, etc.) are available, and what is happening in the future of satellite communications. There are also software programs and smartphone apps to help. Keeping up with future launches and reception of telemetry are also aspects of the myriad of methods of amateur radio satellite communication. Satellite operation is as varied as any other part of the hobby and AMSAT welcomes you to join our membership ranks. Go to **www.amsat.org** and join now.

Since a love of history has been a large part of my experience, I'll leave you with a 1985 quote from Dave Sumner, K1ZZ, then ARRL General Manager, that appeared in a previous satellite book. It was true back then, and it applies to this book today: "You are a part of that reality [space communications]! From setting up a modest ground station and communicating through the 'birds,' to understanding some of the more advanced concepts of satellite orbits and tracking. Whether you are a beginner, an old hand at satellite work, or a student of space science, this book is your launch vehicle into the fascinating journey of amateur radio in space."

I welcome you to pioneer your own out-of-this-world experience!

Joe Spier, K6WAO
President, AMSAT-NA

Acknowledgements

This book would not have been possible without the excellent information provided by AMSAT-NA, particularly their contributions to Chapter 1 concerning the history of the amateur radio satellite program.

I'm also thankful for the contributions of two individuals in particular: AMSAT-NA President Joe Spier, K6WAO, for this gracious foreword, and Rosalie White, K1STO, for her superb discussion of ARISS in Chapter 5.

Chapter 1

The Rich History of Amateur Satellites

Many people are astonished to discover that amateur radio satellites are not new phenomena. In fact, the story of amateur radio satellites is as old as the Space Age itself.

The Space Age is said to have begun on October 4, 1957. That was the day when the Soviet Union shocked the world by launching Sputnik 1, the first artificial satellite. Hams throughout the world monitored Sputnik's telemetry beacons at 20.005 and 40.002 MHz as it orbited the Earth. During Sputnik's 22-day voyage, amateur radio was in the media spotlight since hams were among the only civilian sources of news about the revolutionary spacecraft.

Almost 4 months later, the United States responded with the launch of the Explorer 1 satellite on January 31, 1958. At about that same time, a group of amateur radio operators on the West Coast began considering the possibility of a ham satellite. This group later organized itself as Project OSCAR (Orbiting Satellite Carrying Amateur Radio) with the expressed aim of building and launching amateur satellites. (See the sidebar, "When Does a Satellite Become an OSCAR?")

After a series of high-level exchanges with the American Radio Relay League and the United States Air Force, they secured a launch opportunity. The first amateur radio satellite, known as **OSCAR 1**, would "piggyback" with the Discoverer 36 spacecraft being launched from Vandenberg Air Force Base in California. Both "birds" (as satellites are called among their builders and users) successfully reached low Earth orbit on the morning of December 12, 1961.

OSCAR 1 weighed in at only 10 pounds. It was built, quite literally, in the basements and garages of the Project OSCAR team. It carried a small beacon transmitter that allowed ground stations to measure radio propagation through

The Soviet Union began the Space Age with the launch of Sputnik 1 in 1957.

the ionosphere. The beacon also transmitted telemetry indicating the internal temperature of the satellite.

OSCAR 1 was an overwhelming success. More than 570 amateurs in 28 countries forwarded observations to the Project OSCAR data collection center. OSCAR 1 only lasted 22 days in orbit before burning up as it reentered the atmosphere, but amateur radio's "low tech" entry into the high-tech world of space travel had been firmly secured. When scientific groups asked the Air Force for advice on secondary payloads, the Air Force suggested they study the OSCAR design. What's more, OSCAR 1's bargain-basement procurement approach and management philosophy would become the hallmark of all the OSCAR satellite projects that followed, even to this day.

OSCAR 2 was built by the same team, and although it was similar to OSCAR 1 there were several improvements. One such upgrade modified the internal temperature sensing mechanism for improved accuracy. Another improvement modified the external coating of the satellite to achieve a cooler internal environment. Yet another modification lowered the beacon transmitter output to extend the battery life of the satellite. Thus, the "continuous improvement" strategy that has also become a central part of the amateur satellite approach was set into place very early in the program. On June 2, 1962, it blasted to orbit from Vandenberg Air Force Base in California aboard a Thor Agena B rocket.

OSCAR 2 was followed by **OSCAR 3** on March 9, 1965. OSCAR 3 would become the first amateur radio satellite to carry a *linear transponder* to allow the satellite to act as a communications relay. The transponder was designed to receive a 50 kHz-wide band of uplink signals near 146 MHz and then retransmit them near 144 MHz. This would allow amateurs with relatively modest Earth stations to communicate over much longer distances at these frequencies.

OSCAR 3's transponder operated

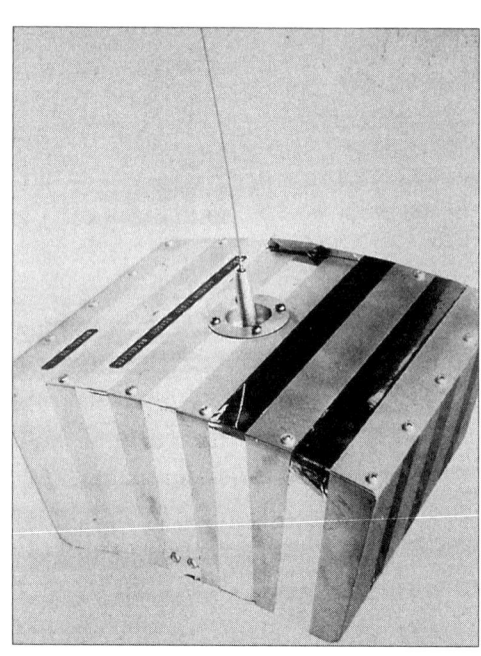

OSCAR 1 was the first amateur radio satellite.

> ## When Does a Satellite Become an OSCAR?
>
> While worldwide AMSAT organizations are largely responsible for the design and construction of the modern-day amateur radio satellites, the original "OSCAR" designation is still being applied to many satellites carrying amateur radio. However, most amateur radio satellites are not usually assigned their sequential OSCAR numbers until *after* they successfully achieve orbit and become operational. Even then, an OSCAR number is only assigned after its sponsor formally requests one.
>
> For example, let's make up a satellite and call it ROVER. The ROVER spacecraft won't receive an OSCAR designation until (1) it reaches orbit and (2) its sponsor submits a request. Now let's presume that ROVER makes it into orbit and the OSCAR request is made and granted. ROVER is now tagged as OSCAR 105 and its full name becomes ROVER-OSCAR 105. You'll find, however, that many hams will abbreviate the nomenclature. Some will simply call the satellite ROVER, or OSCAR 105. They may even abbreviate its full name to just RO-105.
>
> If a satellite subsequently fails in orbit, or it re-enters the Earth's atmosphere, its OSCAR number is usually retired, never to be issued again.

for 18 days, during which time about 1,000 amateurs in 22 countries were heard operating through it. The satellite was the first to clearly demonstrate that multiple stations could successfully use a satellite simultaneously, a technology that is largely taken for granted in satellite telecommunications today.

The fourth amateur radio satellite, **OSCAR 4**, was targeted for a geostationary circular orbit 22,000 miles above the Earth. OSCAR 4 would ride into space aboard a Titan IIIC rocket on December 21, 1965. Unfortunately, despite a valiant effort on the part of the hams and others involved (most of whom were members of the TRW Radio Club of Redondo Beach, California), the top stage of the launch vehicle failed, and OSCAR 4 never reached its intended orbit. Despite this apparently fatal blow, OSCAR 4 operated long enough for amateurs to successfully develop innovative workaround procedures to salvage as much use out of the satellite as possible.

The Birth of AMSAT

In 1969, the **Radio Amateur Satellite Corporation** (AMSAT) was formed in Washington, DC. AMSAT has participated in most amateur satellite projects, both in the United States and internationally. Now, many countries have their own AMSAT organizations, such as AMSAT-UK in England, AMSAT-DL in Germany, AMSAT-BR in Brazil, and AMSAT-LU in Argentina. All these organizations operate independently but may cooperate on large satellite projects and other items of interest to the worldwide amateur radio satellite community. Because of the many AMSAT organizations now in existence, the US AMSAT organization is frequently designated AMSAT-NA (North America).

Since the first OSCAR satellites were launched in the early 1960s, AMSAT's international volunteers have pioneered a wide variety of new communications technologies. These breakthroughs have included some of the first satellite voice transponders as well as highly advanced digital "store-and-forward" messaging transponder techniques. All these accomplishments have been achieved through close cooperation with international space agencies, which often have provided launch opportunities at significantly reduced costs in return for AMSAT's technical assistance in developing new ways to launch paying customers. Such spacecraft design, development, and construction efforts have also occurred in a fiscal environment of individual AMSAT member donations, thousands of hours of volunteer effort, and the creative use of leftover materials donated from aerospace industries worldwide.

AMSAT's major source of operating revenue is obtained by offering yearly or lifetime memberships in the various international AMSAT organizations. Membership is open to radio amateurs and to others interested in the amateur exploration of space. Modest donations are also sought for tracking software and other satellite-related publications at amateur radio gatherings. In addition, specific spacecraft development funds are established from time to time, to receive both individual and corporate donations to help fund major AMSAT spacecraft projects.

From a personnel standpoint, AMSAT-NA is a true volunteer operation. The only person in the entire organization drawing a regular paycheck is the office manager at their headquarters near Washington, DC. She conducts the day-to-day business of membership administration and other key organizational tasks. The rest of AMSAT-NA, from the president of the corporation on down to the workers designing and building space hardware, all donate their time and talents to the organization.

The 1970s

The 1970s was a prolific decade for amateur satellites. The decade began with the launch of **OSCAR 5**. Also known as Australis-OSCAR-5, it was designed and built by students in the Astronautical Society and Radio Club at the University of Melbourne, Australia. OSCAR 5 launched on January 23, 1970, from Vandenberg Air Force Base in California and achieved a 925-mile-high polar orbit aboard a Delta rocket ferrying an American weather satellite.

OSCAR 5 had telemetry beacons transmitting data at 29 and 144 MHz. It was the first amateur satellite to be controlled from the ground; it contained a command receiver that allowed ground stations to control its 29 MHz beacon transmitter. OSCAR 5 did not have a transponder, so it didn't function as a communications relay. However, it did have an innovative magnetic attitude-stabilization system.

OSCAR 6, which reached orbit on October 15, 1972, was the first Phase II satellite. This bird carried a two-way communications transponder that received signals from the ground on 146 MHz and repeated them at 29 MHz with a transmitter power of 1 W.

OSCAR 6 had a sophisticated telemetry beacon that reported information about many parts of the spacecraft, including voltages, currents, and temperatures. OSCAR 6 also had an elaborate ground control system.

Another innovation in OSCAR 6 was *Codestore*, a digital store-and-forward message system. Ground controllers in Canada sent messages to the satellite that were stored and repeated later to ground control stations in Australia.

Static noise plagued OSCAR 6, mimicking signals that the onboard computer interpreted as a command to shut down. To overcome the problem, controllers sent a continuous stream of **ON** commands to the satellite to keep it turned on. The trick worked and OSCAR 6 continued operating for 4.5 years, receiving 80,000 **ON** commands per day!

OSCAR 7, better known as AO-7, was launched on November 15, 1974 by a Delta 2310 booster from Vandenberg Air Force Base in California. OSCAR 7 had two transponders — one received at 146 MHz and repeated what it heard at 29 MHz, while the other listened on 432 MHz and relayed those signals on 146 MHz. The latter transponder, built by radio amateurs in West Germany, had an 8 W downlink transmitter.

In the first satellite-to-satellite link-up in history, a ham transmitted a signal to OSCAR 7, which relayed the signal to OSCAR 6, which then repeated it to a different station on the ground.

"In the first satellite-to-satellite link-up in history, a ham transmitted a signal to OSCAR 7, which relayed the signal to OSCAR 6, which then repeated it to a different station on the ground."

Australians built a telemetry encoder for the satellite and Canadians built a 435 MHz beacon. Other beacons were at 146 and 2304 MHz. The 2304 MHz beacon, with a transmitter power of 100 mW, was built by the San Bernardino Microwave Society of California. Unfortunately, the FCC denied the OSCAR 7 team permission to turn on their 2304 MHz beacon, so it was never tested in space.

OSCAR 7's radio system worked for 6.5 years, until being declared dead in mid-1981 due to battery failure. However, after more than 2 decades of silence, OSCAR 7 came back to life in 2002. Both operating modes still function, but the satellite's control system is not working. Even though it can't be controlled from the ground, OSCAR 7 supports conversations on most daylight passes.

OSCAR 8, the third Phase II amateur satellite, was launched March 5, 1978, on a Delta rocket from Vandenberg Air Force Base in California to a circular 570-mile-high polar orbit. It had two transponders, including one designed by Japanese radio amateurs. It listened at 146 MHz and repeated what it heard through a transmitter on 435 MHz. American, Canadian, and West German amateurs built the rest of OSCAR 8's flight hardware. OSCAR 8 functioned for more than 5 years until its batteries died in 1983.

Russia entered the amateur radio satellite community when an F-2 rocket blasted off on October 26, 1978, from the Northern Cosmodrome at Plesetsk carrying a government satellite and the first two *Radio Sputniks*: **RS-1** and **RS-2**. They were both deployed in an elliptical orbit 1,000 miles above Earth.

Each satellite had a 145 – 29 MHz transponder. The satellites, sometimes referred to as Radio-1 and Radio-2, circled the globe every 120 minutes. They transmitted telemetry beacons in Morse code, relaying temperature and voltage data. These hamsats (ham satellites) had solar cells as well as a Codestore message store-and-forward mailbox. Ground control stations were at Moscow, Novosibirsk, and Arsenyev near Vladivostok.

RS-1 and RS-2 had very sensitive receivers and overload breakers designed to disable the receivers whenever someone used excessive transmitter power on the uplink. (Similar equipment would fly much later with OSCAR 40). The breaker could be reset from the ground when the satellites were over the USSR. However, Western hams, sometimes (needlessly) transmitting with hundreds of watts of power, kept tripping the systems and turning the radio sputniks off. The Russian ground controllers kept resetting the breakers, but most operation ended up being over the Soviet Union since Western hams kept shutting off the transponders when the satellites were over North America and Western Europe.

RS-1 lasted only a few months, but RS-2 was heard until 1981.

The 1980s

Amateur satellite construction and deployment increased dramatically in the 1980s. In fact, 1981 was a record launch year. Eight hamsats blasted to space that year — a tie with 1990 for the most amateur radio satellites launched in a single year.

The '80s saw the debut of the Phase III birds, complex communications satellites designed to be deployed in highly elliptical *Molniya* orbits that take the craft more than 40,000 kilometers into space at the "high" end (apogee) before plunging back toward Earth, skimming a few hundred kilometers above the surface at the closest approach (perigee). At apogee the satellites appear to hover, offering coverage over entire hemispheres for several hours.

The **Phase 3A** satellite was launched on the second flight of Europe's new Ariane rocket on May 23, 1980, from a site outside Kourou, French Guiana, on the northeast coast of South America. Unfortunately, the rocket failed soon after liftoff, sending Phase 3A into the Atlantic Ocean.

Meanwhile, the Russians were busy with projects of their own. Although most Soviet amateur radio satellites were called radio sputniks, there was a series of birds christened *Iskra*, meaning "spark" in Russian. Students and radio amateurs at Moscow's Ordzhjonikidze Aviation Institute built the 62-pound Iskras, each powered by solar cells. Both satellites carried a transponder, telemetry beacon, ground-command radio, Codestore message bulletin board, and computer. The Iskra transponders received at 21 MHz and transmitted at 28 MHz. Their telemetry beacons were near 29 MHz. Controlled by ground stations at Moscow and Kaluga, Iskras were intended for communication among hams in Bulgaria, Cuba, Czechoslovakia, East Germany, Hungary, Laos, Mongolia, Poland, Romania, USSR, and Vietnam.

Iskra-1 was launched July 10, 1981, on an A-1 rocket from the Northern Cosmodrome at Plesetsk to a 400-mile-high polar orbit. After 13 weeks, it burned up while re-entering the atmosphere on October 7, 1981.

The first British amateur radio satellite, **UoSAT-OSCAR 9**, also called UO-9, was designed and built by students at the University of Surrey. In fact, UoSAT is short for University of Surrey Satellite. The 115-pound science and education satellite was blasted to a 340-mile-high polar orbit on October 6, 1981, on a US Delta rocket from Vandenberg Air Force Base in California.

Although OSCAR 9 did not have a communications transponder, it transmitted data and had a television camera that sent pictures back to Earth. The satellite had one of the earliest two-dimensional charge-coupled device (CCD) arrays, forming the first low-cost CCD television camera in orbit. The resulting images transmitted from space were spectacular, considering the early technology. UO-9 was not a stabilized Earth-pointing satellite so the areas covered by its photos were random.

UO-9 had a magnetometer and radiation detectors. In addition, two onboard particle counters measured the effect solar activity and auroras had on radio signals. UO-9 also carried a synthesized radio voicebox with a 150-word vocabulary to announce spacecraft condition reports.

The beacons transmitted at 145 and 435 MHz. For propagation studies, there were additional beacons at shortwave frequencies near 7, 14, 21, and 28 MHz and microwave frequencies near 2 and 10 GHz.

In 1982, a software error mistakenly activated both the 145 and 435 MHz beacons at the same time, preventing the satellite's receiver from hearing command signals from controllers. Surrey hams called on radio amateurs at Stanford University in California to override the jamming. Stanford hams used a 150-foot dish antenna to transmit power equal to 15 mW toward the satellite. The gambit worked and satellite control was later recovered.

After 7 more years of reliable service, OSCAR 9 reentered the atmosphere and was destroyed on October 13, 1989.

In the early 1980s, the amateur radio club at the University of Moscow was busy building a covey of new radio sputniks. Like RS-1 and RS-2, the six satellites each weighed 88 pounds and were housed in cylinders 17 inches in diameter and 15 inches long.

The new birds were launched to 1,000-mile-high orbits on December 17, 1981, on one C-1 rocket from the Northern Cosmodrome at Plesetsk. At that time, it was the largest group of amateur radio satellites ever orbited at one time.

Designated **Radio Sputnik-3 (RS-3) through Radio Sputnik-8 (RS-8)** were in orbits similar to those used by RS-1 and RS-2. The six satellites had transponders receiving at 145 MHz and transmitting at 29 MHz. They also had store-and-forward mailboxes, solar cells, and Morse code temperature and voltage data beacons.

Some of the new radio ssputniks carried the first *autotransponders*. Hams could call a satellite on CW and the robot would respond with a greeting and signal report.

Each RS satellite died as its batteries failed. RS-5 and RS-7 were able to stay on the air until 1988.

The USSR's *Salyut-7* space station was launched to Earth orbit April 19, 1982, with the second 62-lb Iskra satellite inside. Cosmonauts Anatoly Berezovoy and Valentin Lebedev blasted off from Baikonur Cosmodrome on May 13, in a Soyuz transport and docked at *Salyut-7* 2 days later. The cosmonauts unwrapped **Iskra-2** and pushed it out an airlock on May 17, at an altitude of 210 miles.

Iskra-2's telemetry beacon was at 29 MHz. Since it started life in such a low orbit, the satellite was able to remain in space only about 7 weeks before burning up in the atmosphere on July 9, 1982.

Yet another Iskra, **Iskra-3**, was hand-launched from the *Salyut-7* airlock at an altitude of 220 miles on November 18, 1982. Even though Iskra-3 was much like Iskra-2, it suffered from internal overheating and didn't work as well. Iskra-3's telemetry beacon also was at 29 MHz. The third Iskra remained in

An Ariane rocket carried AMSAT-OSCAR 10 to orbit on June 16, 1983.

space only 4 weeks before descending into the atmosphere and burning on December 16.

After the destruction of Phase 3A in 1980, AMSAT immediately began work on Phase 3B. The 200-pound clone was built mostly by German hams and launched on an Ariane rocket on June 16, 1983. It was later named **AMSAT-OSCAR 10**.

Like its predecessor, however, OSCAR 10 had a run of bad luck. Seconds after OSCAR 10's deployment, it was struck by the final stage of the Ariane booster. The collision damaged an antenna and sent the satellite spinning wildly. AMSAT had to wait for the satellite to stabilize before firing an internal thruster to change the orbit on July 11.

The main kick-motor firing did not go well, either. It didn't shut off as ordered and the satellite shot into an exaggerated orbit taking it nearly twice as far away from Earth as planned. Another motor firing was attempted July 26, but, by that time, helium had leaked from the satellite (probably as a result of the Ariane booster collision) which, in turn, most likely caused a failure of the kick-motor's helium-activated fuel valves to operate properly. As a result, OSCAR 10 wound up in an odd orbit ranging from 2,390 to 22,126 miles.

AMSAT had a crippled satellite on its hands. The damaged antenna wouldn't work right, and the orbit exposed the satellite to excessive radiation. The incorrect attitude kept the solar panels from orienting toward the Sun so the batteries couldn't charge properly. AO-10's transponders worked, but the broken antenna and low orbital inclination made it less useful. Its signals were weak and access time was limited. Even so, hundreds of radio amateurs used the satellite.

However, because radiation-hardened computer chips were prohibitively expensive at the time of its construction and launch, AO-10's builders were forced to use non-radiation-hardened computer chips in OSCAR 10's onboard computer. Unfortunately, AO-10's odd orbit was causing it to endure a continuous, intense bombardment of subatomic particles trapped in the Earth's magnetic field. This, in turn, caused a slow destruction of OSCAR 10's non-radiation-hardened computer memory chips. By 1986, AO-10's command computer had deteriorated to the point that mysterious data bits were regularly turning up in the telemetry stream and the satellite was becoming ever harder to stabilize and control. AO-10's transponders would switch off from time to time as voltage dropped when sunlight was low. The satellite required solar illumination 90% of the time, but sometimes received only 50%. When this happened, AO-10 would turn itself off and a command station would be required to transmit a reset order. OSCAR-10 continued this erratic operation until it finally became silent in the 1990s.

The second science and education satellite built by students at England's University of Surrey was UoSAT-B, launched March 2, 1984, from California to a 430-mile-high polar orbit. The 132-pound satellite was eventually renamed **UoSAT-OSCAR 11** (UO-11). It also has been called UoSAT-2.

UO-11's beacons transmit at 145, 435, and 2401 MHz. It handled messages while photographing aurora over the Poles with a sensitive camera which stores the images in memory. Digital telemetry beacons relayed news bulletins from AMSAT and UoSAT, which is headquartered at the Spacecraft Engineering Research Unit at the University of Surrey. When this book was written, OSCAR 11 was operating only occasionally.

Japanese amateurs reached space in 1986, with the launch of the Japan Amateur Satellite (JAS-1a) on August 12 from Japan's Tanegashima Space Center. AMSAT labeled it OSCAR-12; Japanese hams called it Fuji. It came to be known as **Fuji-OSCAR 12** or simply FO-12.

The second science and education satellite built by students at England's University of Surrey was UoSAT-OSCAR 11, launched March 2, 1984 from California to a 430-mile-high polar orbit.

FO-12 had a transponder that received at 145 and retransmitted at 435 MHz. Primarily a packet radio satellite, or *pacsat*, Fuji's transponder could be used either as a message bulletin-board or as a voice repeater. Fuji's telemetry beacon sent data in 20 words per minute (WPM) Morse code.

The mailbox in the sky received typewritten messages from individual ham stations and stored them in a 1.5-MB RAM memory. This electronic message center permitted amateurs to place messages on the satellite's bulletin board to be read by others.

Users were disappointed when FO-12's solar generator was unable to produce sufficient electricity for Fuji's battery. Japanese controllers were forced to turn the satellite off on November 5, 1989.

Soviet hams planned to return to space with **Radio Sputnik-9** in the mid-1980s, but its launch was delayed repeatedly. Finally, the flight was cancelled and the number RS-9 retired permanently.

That disappointment notwithstanding, Soviet hams delighted the amateur satellite world on June 23, 1987, with the launch of a combo package of hamsats, **RS-10 and RS-11**, aboard one large government spacecraft. Radio Sputnik-10/11 went to a 621-mile-high circular orbit as part of the Russian navigation satellite Cosmos 1861, which circles the globe every 105 minutes. The two satellites were, in fact, communications modules riding piggyback on Cosmos 1861 and sharing its considerable electric power budget.

RS-10 and RS-11's telemetry beacons transmitted near 29 and 145 MHz. They had identical shortwave and VHF transponders, but the specific frequencies they used were different. Hams on the ground sent signals to RS-10 and RS-11 on frequencies near 21 and 145 MHz. Downlink signals from the satellites were at 29 and 145 MHz. RS-10 and 11 also featured robot autotransponders that responded to Morse code calls with greetings and contact numbers.

The Rich History of Amateur Satellites   1-9

QSL cards were even issued for contacts with the autotransponders.

RS-10/11 became two of the most popular amateur satellites in history. Anyone with an HF transceiver and the ability to send or receive SSB or CW on 2 meters could access the birds. Simple wire antennas were all that were required. On some weekends, the transponders would be crowded with signals as hams took advantage of the 15-minute contact windows.

When the Russians decommissioned Cosmos 1861 in the 1990s, that spelled the end of RS-10/11 as well.

"Simple wire antennas were all that were required. On some weekends, the transponders would be crowded with signals as hams took advantage of the 15-minute contact windows."

As the saying goes, the third time is the charm. That was certainly the case for AMSAT's third attempt at a Phase III satellite. **AMSAT-OSCAR 13**, which safely reached orbit on June 15, 1988. In its Molniya orbit, OSCAR 13 reached and apogee of 22,000 miles before sweeping back around the Earth at an altitude of 1,500 miles.

OSCAR 13 was the most complex Phase III satellite of its time. It was a project of AMSAT-DL. OSCAR 13 offered four transponders for packet, facsimile, slow-scan television, voice (SSB), radioteletype (RTTY), and Morse code (CW). Transponders received at 435 and 1269 MHz and retransmitted at 145, 435, and 2400 MHz. The satellite's computer followed a schedule to manage transponder activation and use.

With its near-hemispheric coverage at apogee, OSCAR 13 quickly became famous as a "DX satellite." Hams in different continents could communicate for hours at a time. There were even scheduled roundtable chats via OSCAR 13.

All good things must come to an end, and so did OSCAR 13. The satellite re-entered Earth's atmosphere on December 6, 1996.

The Last Satellites of the 20th Century

With the dawn of the 1990s, space enthusiasts at England's University of Surrey had several new satellites ready to fly. At about the same time, another international group of AMSAT experimenters had devised a new, far more radical satellite design approach to take advantage of rapid advances in solar cell efficiency. Rather than using bulky spaceframes, these new birds were small boxes only 22.6 × 22.6 × 22.3 centimeters on each side and weighing only 13 kilograms. Known as *Microsats*, these satellites represented a pioneering approach to satellite design, one that commercial satellite builders would soon follow.

On January 22, 1990, an entire fleet of Microsats (along with two new UoSATs) were placed in orbit by a single Ariane 4 rocket. The lineup included **UoSAT-OSCAR 14**, **UoSAT-OSCAR 15**, **AMSAT-OSCAR 16**, **DOVE-OSCAR 17**, **WEBERSAT-OSCAR 18**, and **LUSAT-OSCAR 19**.

UoSAT-OSCAR 14 spent its first 18 months in orbit operating as a packet radio store-and-forward satellite. It received electronic mail and stored it for later reading by ground stations in other parts of the globe. In early 1992, all amateur operations were moved from AO-14 to UoSAT-OSCAR 22. AO-14 operations were then dedicated for use by Volunteers In Technical Assistance (VITA) who used it for sending and receiving email messages in Africa.

Several years later, the computer used for store-and-forward communications became non-operational. In March 2000, UO-14 was returned to amateur use and reconfigured as a single-channel FM repeater. This move dramatically increased its popularity in the ham community. For the first time, hams were able to enjoy satellite communication using nothing more than common dual-band FM transceivers.

OSCAR 14 remained in service until November 2003, when it was officially declared dead. The Mission Control Centre at the Surrey Satellite Technology Ltd reported that the venerable and popular bird "had reached the end of its mission after nearly 14 years in orbit."

OSCAR 15 made it to orbit with its companions, but its lifetime was short. Within hours of being deployed, it fell silent.

AMSAT-OSCAR 16 was designed to be a dedicated store-and-forward file server in space. Using 1200 bit per second radio links, AMSAT-OSCAR 16 interacted with ground station terminal software to appear as a packet radio bulletin board system to the user. Anyone wishing to download files and personal email from anywhere in the world could request that information be "broadcast" to everyone under the footprint of the spacecraft or directed specifically to a particular ground station. Popular as it was, OSCAR 16 finally fell silent a few years later.

DOVE-OSCAR 17 was sponsored by AMSAT-BR (Brazil). The project was led by Dr. Junior Torres de Castro, PY2BJO. DOVE, an acronym for Digital Orbiting Voice Encoder, carried hardware capable of reproducing digitized speech, or controlling a Votrax speech synthesizer. However, due to hardware failures that occurred after launch, the primary mission of broadcasting voice messages of world peace was not fully realized.

DOVE operated sporadically on a downlink frequency of 145.825 MHz FM, transmitting AX.25 protocol packet radio telemetry. Using 1,200 bps Bell 202 style AFSK emissions, DOVE-OSCAR 17 could be copied with packet radio equipment in wide use on VHF at the time. Today, DOVE is no longer operational.

Sporting a full-color charge coupled device (CCD) camera, **WEBERSAT-OSCAR 18** digitized Earth images and downlinked them as an AX.25 serial data stream. WEBERSAT-OSCAR 18 was a product of the efforts of the Center for Aerospace Technology (CAST) at Weber State University in Ogden, Utah. WO-18's CCD camera had a resolution of 700 pixels by 400 lines and could be viewed with *Weberware* software running on a personal computer having adequate graphics display capability. Digitized National Television Standard Committee (NTSC) video from the camera was assembled into packets that were sent as unnumbered information (UI) frames. Ground stations had to receive this data over several passes to capture a complete image. Each image contained about 200 kb of data.

The final member of the 1990 Microsat fleet was **LUSAT-OSCAR 19**. OSCAR 19 is coordinated by AMSAT-LU (Argentina) and is a packet radio store-and-forward spacecraft, much the same as AMSAT-OSCAR 16. The only difference

The LUSAT-OSCAR 19 satellite is typical of Microsat design.

In March 2000, UO-14 was configured as a single-channel FM repeater. It remained in service until 2003.

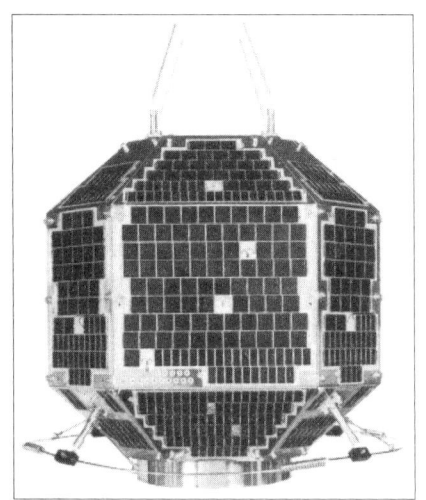

The Japanese launched Fuji-OSCAR 20 on February 7, 1990, from the Tanegashima Space Center on an H-1 two-stage rocket.

between the two satellites is that AO-16 supports an S-band beacon in addition to the mailbox, while LO-19 has a 70-centimeter CW beacon. Today, only the CW beacon remains operational.

In the wake of the Microsat successes, the Japanese launched their next amateur radio satellite, **Fuji-OSCAR 20**, on February 7, 1990, from the Tanegashima Space Center on an H-1 two-stage rocket. Also known as FO-20, its orbit differed slightly from most OSCAR satellites, being slightly elliptical with a high inclination. This assured that the satellite would remain in sunlight for most of its orbit.

The physical structure of FO-20 was that of a 26-sided polyhedron with a weight of approximately 50 kilograms, so it was much larger than the Microsats. Although Fuji-OSCAR 20 used AX.25 packet radio communications links as the Microsats did, one big difference between FO-20 and the Microsats was that FO-20's packet radio could be accessed without the need for special Microsat terminal software. Any computer or terminal that could be used to access terrestrial packet radio bulletin board systems (BBSs) could be used to access the FO-20 mailbox.

The other difference between FO-20 and the Microsat satellites was that, in addition to the packet mailbox features of the satellite, FO-20 also supported an analog linear transponder for SSB and CW communications. FO-20 suffered from a declining power generation capacity over the years and eventually became silent.

The year 1990 would see one more amateur satellite launch. It would also be the first from Pakistan. Known as **Badr-1**, the satellite was launched July 16, 1990 by China, on one of its Long March rockets to a 375-mile-high circular orbit. The 150-pound Badr-1 was constructed by engineers who were hams at the Space and Upper Atmosphere Research Commission (SUPARCO) at the University of the Punjab at Lahore. Several had completed master's degrees in engineering at England's University of Surrey. Back home, they used their new knowledge to build the satellite with support from the Pakistan Amateur Radio Society.

Badr-1's glory was brief. It reentered the atmosphere on December 9, 1990.

Radio Sputnik-14/AMSAT-OSCAR 21 was the first satellite of 1991, taking to the skies on January 29, 1991, as a joint venture between AMSAT-U (Russia) and AMSAT-DL (Germany). It was essentially a module riding inside a larger Russian government satellite known as INFORMATOR-1. Amateur radio lost a valuable asset when the Russian government ran out of funds for the project and turned the entire satellite off on September 16, 1994. RS-14/AO-21 had been a popular satellite because it was easy to use. The hamsat was an FM repeater that also transmitted recorded messages commemorating events like the 25th anniversary of the first landing on the Moon.

Another radio sputnik, **RS-12/13**, was launched on February 5, 1991. Like the popular RS-10/11 combo, RS-12/13 module rode piggyback on a Russian COSMOS navigation satellite. Each radio sputnik had a 40 kHz-wide linear transponder allowing for many simultaneous CW and SSB contacts. They also carried autotransponders that acknowledged CW calls with a greeting and contact number.

RS-12/13 was quite popular because of its HF and low VHF transponders. Anyone with simple station equipment could work them. The satellites enjoyed an 11-year lifespan until a solar flare disabled the parent COSMOS satellite in August 2002.

The British Microsat UoSAT-5, built by the University of Surrey, was launched July 17, 1991, to become **UoSAT-OSCAR 22**. OSCAR 22 was designed to serve several missions. One mission was to carry out experiments originally slated for UoSAT-OSCAR 15, whose on-board electronics failed shortly after the spacecraft reached orbit. The primary purpose of UoSAT-OSCAR 22, however, was to provide non-amateur radio related store-and-forward digital communications for the non-profit, humanitarian organizations VITA and SatelLife.

UoSAT-OSCAR 22's primary role in space was later modified when the satellite played a part in amateur radio's first "role reversal" with another spacecraft, UoSAT-OSCAR 14. One of the most unique aspects of UoSAT-OSCAR 22 was its Earth Imaging System (EIS), designed by University of Surrey doctoral student Marc Fouquet. The Earth System was designed to capture Earth images from low orbit using a CCD camera and broadcast those images to ground stations using amateur frequencies and the AX.25 packet radio "Pacsat Broadcast Protocol." After years of reliable service, UO-22 fell silent.

The year 1992 saw the launch of **KITSAT-OSCAR 23** on August 10. It was a Microsat designed by amateurs from the Korean Advanced Institute of Science and Technology (KAIST) studying at the University of Surrey in the UK. The digital store-and-forward satellite was managed by the KAIST Satellite Technology Research Center (SaTReC) of South Korea. It enjoyed several years of life before its power systems eventually failed.

The first French amateur satellite, **Arsene-OSCAR 24**, flew on May 13, 1993, aboard an Ariane V-56A from Kourou, French Guiana. Built by the Radio Amateur Club de l'Espace, the satellite had a short, troubled life. It was originally intended to be a packet relay satellite, but the packet system was never implemented because the 2-meter transponder failed soon after launch. Arsene was then used to relay SSB and CW signals on 2.4 GHz for several months until this transponder failed as well.

There was another amateur satellite "first" in 1993. Launched alongside HealthSat-2 in September 1993, **PoSAT-1** was Portugal's first satellite. PoSAT-1 was built at the University of Surrey in a collaborative program between a consortium of Portuguese academia and industry. The satellite carried several technology experiments and had the ability to function as a packet radio store-and-forward system.

PoSAT-1 operated on amateur frequencies for several weeks in early 1994. *OSCAR News* (February 1994, p. 35) carried a letter from CT1DBS reporting that an agreement had been signed by AMSAT-PO and the PoSAT Consortium on December 6, 1993, stating, "The name of PoSat-1, when in use by the amateur radio community will be **PoSAT-OSCAR 28**, OSCAR 28 or PO-28." Despite having an OSCAR designation, PoSAT-1 was never re-opened for amateur operations.

KITSAT-OSCAR 25 was launched September 26, 1993, by an Ariane V59 rocket from Kourou, French Guiana. This satellite was essentially a twin of KITSAT-OSCAR 23, but it was designed and built entirely by KAIST. KO-25 was operated from SaTReC in South Korea. KO-25's mission was to take CCD pictures, process numerical information, measure radiation, and receive and forward messages. The Infrared Sensor Experiment (IREX) was designed to measure the characteristics of infrared sensors in space. A passive cooling structure was also devised for this experiment so ground controllers could monitor its temperature. In its amateur radio role, KO-25 functioned as a packet store-and-forward relay, although it is presently nonoperational.

The same rocket that took KO-25 to orbit also carried Microsats **ITAMSAT-OSCAR 26** and **AMRAD-OSCAR 27**.

ITAMSAT was the first Italian amateur satellite. It was built and operated by a small team of AMSAT members from Italy. Its mission was to store and forward amateur radio messages in the same manner as AO-16, LO-19, UO-22, KO-23, and KO-25. Unfortunately, OSCAR 26 saw little ham use and is now silent.

The satellite that would eventually become OSCAR 27 undergoing final checkout.

In contrast to OSCAR 26, AMRAD-OSCAR 27 (AO-27) was highly popular and remains so today. AO-27 is a secondary amateur communications payload carried aboard the EYESAT-1 experimental Microsat built by Interferometrics Incorporated of Chantilly, Virginia. The commercial side of the spacecraft's mission is the experimental monitoring of mobile industrial equipment.

The amateur radio portion of the satellite was constructed by members of AMRAD, a technically oriented, non-profit organization of radio amateurs based in the Virginia suburbs of Washington, DC. It was intended to be a platform to conduct digital satellite communications experiments.

The only amateur satellite of 1994 was **Radio Sputnik 15**. It was launched on December 16. It carried a 2 – 10 meter transponder and CW beacons, but their signals were so weak that it is doubtful the satellite ever became fully operational. Hams today occasionally report reception of weak CW beacons from RS-15.

On March 28, 1995, the Mexican **UNAMSAT-1** and Israeli **TechSat-1** amateur satellites were launched from Russia's Plesetsk Cosmodrome aboard a Start-1 rocket. The rocket exploded soon after liftoff, destroying both satellites.

The Japanese **Fuji-OSCAR 29**, the successor to the failing OSCAR 20, reached orbit on August 17, 1996, from the Tanegashima Space Center aboard a H-II No. 4 rocket. It was a project of the Japan Amateur Radio League. OSCAR 29 carried an SSB/CW analog linear transponder that listened on 2 meters and repeated on 70 centimeters. It enjoyed 11 years of life before being declared dead in 2007.

The second Mexican hamsat, UNAMSAT-B, built at the National University of Mexico (UNAM)), was launched from Russia on September 5, 1996. **Mexi-**

co-OSCAR 30, as it came to be known, was the twin of UNAMSAT-1 (see above). Unfortunately, MO-30 stopped working when its receiver failed within a few hours after reaching orbit.

Radio Sputnik 16 took to the skies in February 1997, carrying 2 – 10-meter transponders and a 70 centimeter beacon. For unknown reasons, however, the transponders were never activated. The satellite reentered the atmosphere in 1999.

Radio Sputnik 16 was a satellite shrouded in mystery. It was launched successfully in 1997, but its transponders were never activated.

An unusual satellite made a brief appearance on November 4, 1997. **Radio Sputnik 17a**, also known as **Sputnik 40**, was hand-launched from the Russian *Mir* space station. The little satellite was a scale model built by high school students to commemorate the 40th anniversary of the launch of Sputnik 1. RS-17a broadcasted a signal for 55 days and was last heard on December 29, 1997.

Thai-Microsatellite-OSCAR 31 was launched on July 10, 1998, from the Russian Baikonur Cosmodrome into a circular sun-synchronous orbit. It carried a 9600-baud FSK digital transponder, GPS receiver and imaging subsystem. OSCAR 31 was like KITSAT-OSCAR 23, but included the ability to take multispectral images. At the time of this writing, it is nonoperational.

Israel's **Gurwin-OSCAR 32**, otherwise known as TechSat-1b, soared into orbit on July 10, 1998. The Microsat was a project of students and scientists at the Technion Institute of Technology in Haifa, along with the Israel Amateur Radio Club. Eight years after launch, OSCAR 32 is still going strong. The satellite supports 9600-baud store-and-forward packet and a high-resolution color camera supplying stunning images of the Earth. In recent years, OSCAR 32 has also been used as a 9600 baud relay for the Automatic Position Reporting System (APRS) with a dedicated uplink on 145.930 MHz.

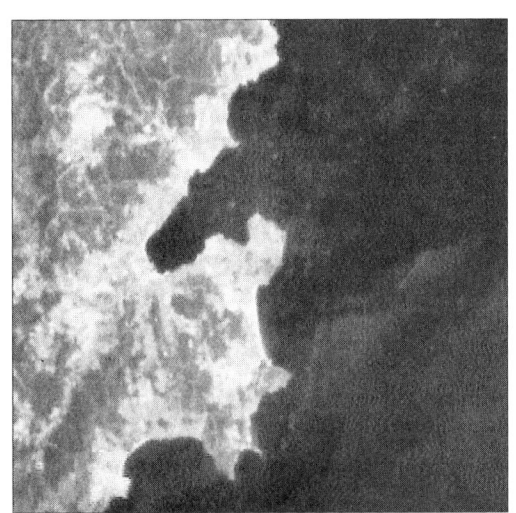

The first image transmitted from Gurwin-OSCAR 32.

The University of Alabama in Huntsville, in conjunction with Students for the Exploration and Development of Space (SEDS) USA, designed and constructed SEDSAT-1, or **SEDSat-OSCAR 33** as it came to be known. It reached orbit on a Delta II rocket from Vandenberg Air Force Base in California on October 24, 1998. It carried a packet store-and-forward transponder, an analog transponder, and several experiments. It is currently only semi-operational.

Launched October 30, 1998 from the space shuttle *Discovery*, **PANSAT-OSCAR 34** was a Microsat designed to provide a packet store-and-forward message system. Designed and built at the Naval Postgraduate School in Monterey, California, OSCAR 34 was unique

The Rich History of Amateur Satellites   1-15

among the digital satellites in that it employed direct sequence spread-spectrum communications. OSCAR 34 was only available for limited amateur access, however, and finally became silent.

Another tiny radio sputnik, this one called **Radio Sputnik-18/Sputnik-41**, was tossed out of the *Mir* space station airlock on November 10, 1998. Like its predecessor, Sputnik 40, it was a ⅓-scale replica of Sputnik 1. It transmitted a beacon signal on 2 meters along with voice greetings in English, Russian, and French. The satellite reentered the atmosphere after a few months in orbit.

SUNSAT-OSCAR 35 was a Microsat built by post-graduate engineering students in the Electronic Systems Laboratory of the Department of Electrical and Electronic Engineering at the University of Stellenbosch in Matieland, South Africa. It was launched on February 23, 1999. Payloads included NASA experiments, amateur radio communications, a high-resolution imager, precision attitude control, and school experiments. OSCAR 35 became silent in February 2001.

Yet a third satellite was hand-launched from the *Mir* space station on April 16, 1999. **Radio Sputnik-19/Sputnik-99** was controversial because it was originally designed to broadcast a beacon on ham frequencies that promoted the Swatch watch company of Switzerland. After substantial outcry from the ham community, the amateur radio element of the project was abandoned, and the satellite was launched with its ham transmitter turned off.

The last amateur radio satellite of the 20th century was **UoSAT-OSCAR 36** built by Surrey Satellite Technology Ltd at the University of Surrey in the United Kingdom. It was launched on April 21, 1999, on a Russian rocket from the Baikonur Cosmodrome. OSCAR 36 carried a number of imaging payloads and a unique propulsion system for orbital housekeeping experiments. The S-band downlink ran at speeds up to 1 Mb/s for downloading imaging data. For amateurs, it functioned as a packet store-and-forward satellite during its brief functional lifetime.

The Dawn of a New Century

A unique rocket and a complex set of amateur satellites lifted off less than a month into the 21st century. On January 27, 2000, a Minotaur rocket launched from Vandenberg Air Force Base in California. The six-story Minotaur was built from parts recycled from retired nuclear missiles. It combined the first and second stages of a decommissioned US Air Force Minuteman-2 missile with the third and fourth stages of an Orbital Sciences' commercial air-launched Pegasus rocket. The launch proved that the combo was capable of ferrying satellites to space.

The Minotaur carried several non-ham satellites to orbit, along with a group of amateur radio birds:

Arizona State-OSCAR 37 contained an amateur packet hardware system and a 2-meter/70-centimeter FM voice repeater. The satellite successfully activated, but telemetry soon confirmed that a critical problem had occurred in the power system. The solar arrays were offline, and the batteries could not be recharged. As a result, OSCAR 37 died 15 hours later.

OPAL-OSCAR 38 carried a 9600-baud packet radio store-and-forward system. It is now nonoperational. OSCAR 38 was designed to launch six minia-

OPAL-OSCAR 38 carried a 9600-baud packet radio store-and-forward system and launched several tiny satellites, known as *picosats*.

ture satellites, known as *picosats*. One of these picosats was an amateur radio project known as **StenSAT**. StenSAT was built by amateur radio operators in Washington, DC and was a mere 12 cubic inches in size, and weighing only 8.2 ounces. It featured a single-channel FM voice repeater with uplink at 145.84 MHz and a downlink at 436.625 MHz. Periodically, StenSAT would transmit packets of telemetry. Although StenSAT was successfully deployed, it never became fully operational.

Weber-OSCAR 39, also known as JAWSAT, was designed to serve as a platform for deploying smaller satellites. JAWSAT stands for Joint Air Force Weber Satellite, which was a joint project between the US Air Force and Weber State University. During this mission, JAWSAT deployed OSCAR 37; OSCAR 38; OCSE, the US Air Force Research Laboratory's Optical Calibration Sphere Experiment, and FalconSat. The latter satellite was developed by US Air Force Academy cadets to study how charged particles can build up and then wreak havoc with satellites' onboard computer systems. The telemetry from JAWSAT was transmitted on ham frequencies.

Launched September 26, 2000 aboard a converted Soviet ballistic missile from the Baikonur Cosmodrome, **SaudiSat-OSCAR 41** was one of three amateur radio satellites on the same launch vehicle. OSCAR 41 was one of the first Saudi Arabian Microsats with amateur radio capability. It contained a 9600-baud store-and-forward packet system as well as an analog FM repeater. It was built by the Space Research Institute at the King Abdulaziz City for Science and Technology. Initially, the FM repeater was most active, but the satellite fell silent in February 2003 and has not been heard since.

Accompanying SaudiSat-OSCAR 41 on the flight were **Malaysian-OSCAR 46** and **Saudi-OSCAR 42**. OSCAR 46 was Malaysia's first Microsat. In addition to commercial land and weather imaging payloads, it offered FM and FSK amateur radio communication. It was built as a collaborative effort between the Malaysian government and Surrey Satellite Technology Ltd. Opened briefly for ham use, it is now nonoperational. OSCAR 42, like OSCAR 41, was a Saudi Arabian Microsat. As far as its amateur radio capability was concerned, it was a virtual twin of OSCAR 41. It is now off the air.

The largest, most complex amateur radio satellite ever constructed was launched on an Ariane 5 rocket from Kourou, French Guiana, on November 16, 2000 — **AMSAT-OSCAR 40**, also known as AO-40. Like its Phase III predecessors, AO-40 was launched into an eccentric Molniya orbit ranging from 58,971 km at apogee and 1,000 kilometers at perigee.

OSCAR 40 offered linear transponders on several bands from VHF to microwave, a high-resolution color camera, digital transponders, and scientific experiments. With its high altitude at apogee, powerful output, and gain antennas, it promised hemispheric communications at signal levels never experienced before.

OSCAR 40 was the largest, most complex amateur satellite ever constructed.

Tragically, a protective plastic cap over a vent on the satellite's orbital insertion motor was inadvertently left in place at launch. And while precise details of exactly what happened next will forever remain unclear, it appears that, as a direct result of that inadvertent error, during successive firing attempts of the motor in space, pressure built up in the motor's feed lines to the point that one or more of them eventually ruptured, spewing volatile hypergolic (i.e. "no spark needed") fuel and oxidizer into the interior of the spacecraft. The resulting explosion was apparently strong enough to blow the omni-antenna-laden bottom out of the satellite. All radio links and beacons immediately went silent. Communications were re-established weeks later and the satellite was eventually returned to limited operation. Although a shadow of its former self, hams around the world still enjoyed considerable use of OSCAR 40 until January 2004, when it suffered a catastrophic failure of the main battery. The satellite has been silent ever since and is now considered lost.

On September 30, 2001, an Athena I rocket blasted off from the Kodiak Launch Complex on Kodiak Island, Alaska carrying three amateur radio satellites. **Navy-OSCAR 44** (NO-44), also known as PCSAT-1, is a 1200-baud APRS digipeater designed for use by stations using handheld or mobile transceivers.

The Athena rocket also deployed **Starshine-OSCAR 43**, a basketball-shaped satellite covered with 1,500 aluminum mirrors polished by an estimated 40,000 student volunteers in the United States and 25 other countries. Starshine's primary mission was to involve and educate school children from around the world in space and radio sciences. In addition to helping build Starshine, students were able to visually track the satellite during morning and evening passes by recording its telltale mirror flashes and reporting their observations to Project Starshine headquarters. Starshine-OSCAR 43 transmitted its telemetry on ham frequencies but is now silent.

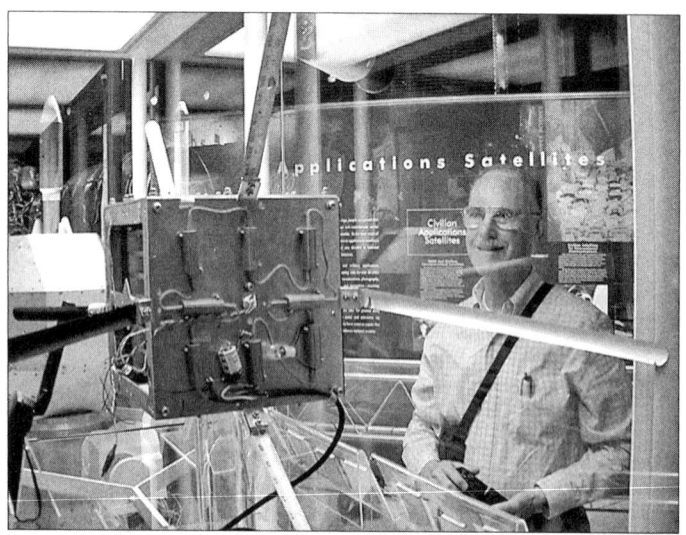

This is a flight model of Navy-OSCAR 44 (NO-44), also known as PCSAT-1. It is a 1200-baud APRS digipeater designed for use by stations using handheld or mobile transceivers. [Courtesy of the SETI LEAGUE]

The last passenger on the Kodiak launch was **Navy-OSCAR 45**, otherwise known as Sapphire. Sapphire was a

Starshine-OSCAR 43 was covered with 1,500 aluminum mirrors polished by an estimated 40,000 student volunteers in the United States and 25 other countries.

Microsat designed and built by students at Stanford University and Washington University in St Louis. The primary mission of Sapphire was to space-qualify two sets of "Tunneling Horizon Detector" infrared sensors designed and built by the Jet Propulsion Laboratory and Stanford University. Secondary experiments included a digital camera and voice synthesizer. Today, OSCAR 45 is nonoperational.

Radio Sputnik 21/Kolibri had a brief, but interesting life. The tiny educational satellite was built by the Special Workshop of Space Research Institute of the Russian Academy of Sciences in Tarusa, Kaluga. Participating students in Sydney, Australia and Obninsk, Russia named it Kolibri-2000. (Kolibri means "hummingbird.")

The satellite rode the Russian *Progress M-17* cargo freighter to the International Space Station in December 2001. The engineers had attached a radio-controlled remote launching port to the Progress capsule that held Kolibri while docked at the ISS. Then, on March 20, 2002, as the Progress rocket departed, the 44-pound Kolibri was ejected into space. Kolibri dropped slowly, circled Earth 711 times, and burned up in the atmosphere after about 4 months. It sent telemetry data and digitally recorded voice messages on the downlink frequency of 145.825 MHz.

French amateurs returned to space in 2002 with two picosats known before launch as IDEFIX CU1 and IDEFIX CU2. They flew into orbit on May 3 on Ariane-4 flight V151 from Kourou, French Guiana. The Ariane also ferried the SPOT-5 photo satellite to orbit. The satellites were later officially named **BreizhSAT-OSCAR 47** and **BreizhSAT-OSCAR 48**. The satellites remained attached to the Ariane rocket third stage and transmitted pre-recorded narrowband FM (NBFM) voice messages and digital telemetry data. Both battery-powered birds became silent two weeks later.

Radio Sputnik 21/Kolibri rode a Russian Progress M-17 cargo freighter to the International Space Station in December 2001. On March 20, 2002, as the Progress rocket backed away and departed the station, the 44-pound Kolibri was ejected into space.

Two ham satellites lifted off on December 20, 2002, on a converted Soviet ballistic missile from the Baikonur Cosmodrome. **AATiS-OSCAR 49** was a German amateur radio payload onboard the small German scientific satellite RUBIN-2. It was designed as a store-and-broadcast system for APRS but failed about a month after reaching orbit. Its launch companion, **Saudi-OSCAR 50**, remains operational today as an orbiting FM repeater.

It is fair to say that 2003 marked the debut of the *CubeSats*. As the name implies, CubeSats are tiny cube-shaped satellites dedicated to specific missions in low Earth orbit, some with short design lifespans. Between 2003 and the present day, a horde of CubeSats have been placed in orbit, but only a few have received OSCAR designations.

The Rich History of Amateur Satellites   1-19

A CubeSat is aptly named, although sometimes their shapes can be more rectangular.

CubeSat-OSCAR 55 (known as Cute-1) and **CubeSat-OSCAR 57** were launched on June 30, 2003, from Baikonur Cosmodrome aboard a Dnepr rocket. Both were University of Tokyo projects and transmitted telemetry on ham frequencies.

On the same day, three CubeSats with ham capability were launched from Plesetsk MSC aboard a Rockot booster. **CanX-1**, **DTUSat**, and **AAU Cubesat** have since reentered the atmosphere.

RS-22, the second radio sputnik of the 21st century, was launched on September 27, 2003, from Baikonur Cosmodrome aboard a Dnepr rocket. It was a project of Mozhaysky Military Space Academy.

The only amateur radio satellite of 2004 turned out to be one of the most popular hamsats of the new century. **AMSAT-OSCAR 51**, also known as Echo, was launched on June 28, 2004, from Baikonur Cosmodrome aboard a Dnepr rocket. AO-51 contained an FM repeater with 144 MHz and 1.2 GHz uplinks and 435 MHz and 2.4 GHz downlinks. It is no longer active.

The Indian **VUSat-OSCAR 52** was launched on May 5, 2005, from Sriharikota, India aboard a PSLV rocket. For several years, it was the most popular satellite for SSB and CW operating. The satellite listened for signals on the 70-centimeter band and repeated what it heard on the 2-meter band. Rather than listening and repeating on a single frequency, however, the OSCAR 52 linear transponder relayed signals across an entire range of frequencies, allowing it to carry many conversations at once. After years of service, it finally became inactive.

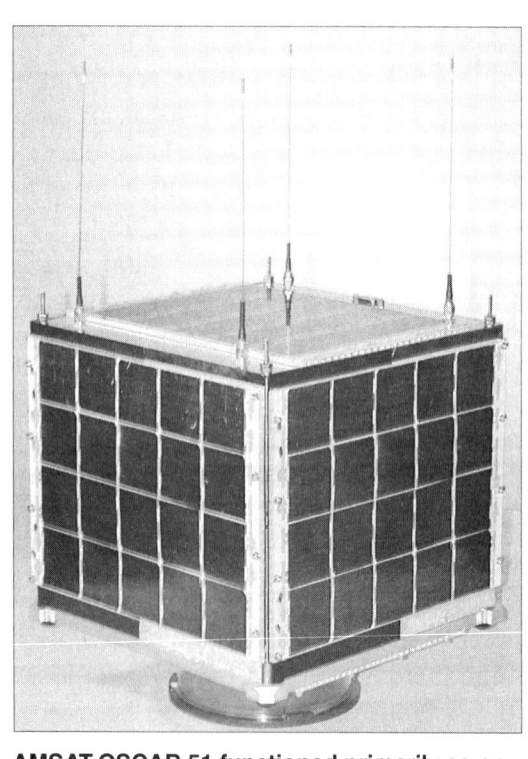

AMSAT-OSCAR 51 functioned primarily as an FM repeater in space for many years.

PCSAT 2 was a "satellite" that resembled a suitcase. It was installed by astronaut Soichi Noguchi on the outside of the International Space Station on August 3, 2005. PCSAT 2 completed a 4-week test of its PSK-31 transponder mode and operated in APRS packet digipeater mode as well. It was returned to Earth on space shuttle mission STS-115 on September 21, 2006.

AMSAT-OSCAR 54, better known as *SuitSat*, is one of the most unusual amateur radio satellites ever placed in orbit. SuitSat was a payload installed inside a discarded Russian Orlan EVA suit that was ejected from the International Space Station on September 8, 2005. It carried a 2-meter amateur radio beacon transmitter. Unfortunately, there was an apparent malfunction of the transmitter or the antenna (or both) and the beacon became very weak, copyable only by hams with well-equipped stations. The last confirmed reception of SuitSat was on Saturday February 18, 2006, by Bob King, VE6BLD.

eXpress-OSCAR 53, known as SSETI Express, was launched on October 27, 2005, from the Plesetsk

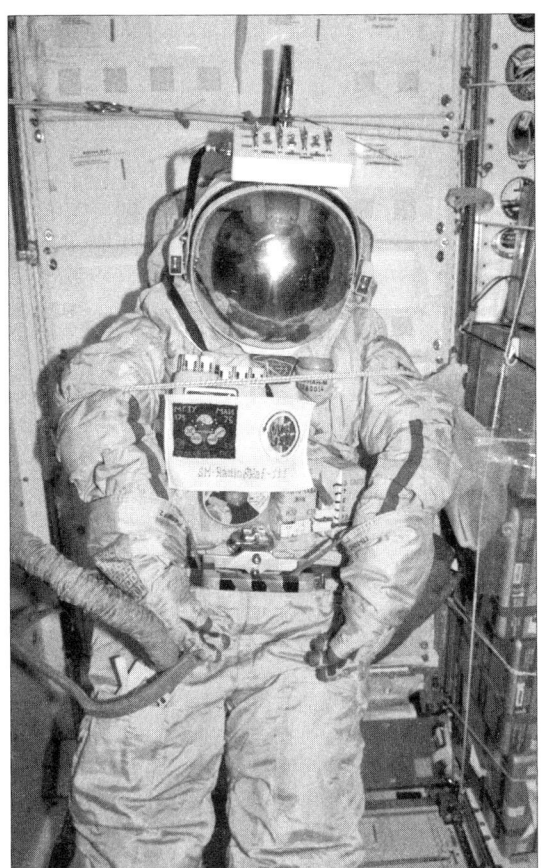

AMSAT-OSCAR 54, better known as SuitSat, was one of the most unusual amateur radio satellites ever placed in orbit. SuitSat was a payload installed in a discarded Russian Orlan EVA suit that was ejected from the International Space Station on September 8, 2005.

Cosmodrome in Russia. Shortly after reaching orbit, SSETI entered safe mode and began to send 9600 baud and carrier pulse telemetry. OSCAR 53 apparently deployed its three research CubeSats but fell silent thereafter.

The Japanese **CubeSat-OSCAR 56** was launched on February 21, 2006, from Kagoshima Space Center aboard a JAXA M-V 8 rocket. It was a project of Tokyo Institute of Technology Matunaga Laboratory for Space Systems (LSS).

Another Japanese CubeSat, **HITSat-OSCAR 59**, was launched on September 22, 2006, from Kagoshima Space Center aboard a JAXA M-V rocket. It features a 1200-baud packet radio bulletin board system and remains operational today.

Navy-OSCAR 61 was carried aboard space shuttle mission STS-116. It functioned as an APRS relay until it reentered the atmosphere on December 25, 2007.

Ten satellites reached orbit April 28, 2008, aboard an Indian PSLV-C9 rocket launched from the Satish Dhawan Space Centre. The primary payloads were India's CARTOSAT-2A and IMS-1 satellites. In addition to the NLS-5 and RUBIN-8 satellites, the rocket carried six CubeSat research satellites, all of which communicate using amateur radio frequencies:

The **SEEDS** satellite was designed and built by students at Japan's Nihon University. SEEDS downloads telemetry in Morse code and 1200-baud FM AFSK packet radio at 437.485 MHz.

AAUSAT-II was the creation of a student team at Aalborg University in Denmark. It downlinked scientific telemetry at 437.425 MHz using 1200 or 9600-baud packet.

Can-X2 was a product of students at the University of Toronto Institute for Aerospace Studies, Space Flight Laboratory (UTIAS/SFL). Can-X2 downlinked telemetry at 437.478 MHz using 4 kbps GFSK.

Compass-One was designed and built by students at RWTH Aachen University of Applied Sciences in Germany. The satellite featured a Morse code telemetry beacon at 437.275 MHz. Compass-1 also provided a packet radio data downlink, which included image data, at 437.405 MHz.

CUTE-1.7 + APDII is a satellite created by students at the Tokyo Institute of Technology. This satellite not only provided telemetry, it also offered a 9600-baud packet store-and-forward message relay with an uplink at 1267.6 MHz and a downlink at 437.475 MHz.

Delfi-C³ was designed and built by students at Delft University of Technology in the Netherlands. It included an SSB/CW linear transponder. Delfi-C3 downlinked 1200-baud packet telemetry at 145.870 MHz. The linear transponder had an uplink passband from 435.530 to 435.570 MHz and a corresponding downlink passband from 145.880 to 145.920 MHz.

Amateur Satellites Today

The decade that followed saw the launches of many more satellites. AMSAT placed several more satellites in orbit, many with FM repeater capabilities. Other nations have also launched satellites that include amateur radio functionality. Among the newest satellites, there are some that do not allow amateurs to communicate through them, but they *do* transmit telemetry that hams can receive and decode. As you'll learn later in this book, these satellites can be used as excellent teaching tools.

Just look at the number of satellites listed below that reached orbit in the last 10 years…

FASTRAC-OSCAR 69
    Launched: 11/20/2010
    Responsible Organization: University of Texas at Austin
FASTRAC-OSCAR 70
    Launched: 11/20/2010
    Responsible Organization: University of Texas at Austin
AubieSat-OSCAR 71
    Launched: 10/28/2011
    Responsible Organization: Auburn University
MaSat-OSCAR 72
    Launched: 2/13/2012
    Responsible Organization: Budapest University of Technology and Economics Hungary
AMSAT-OSCAR 73
    Launched: 11/21/2013
    Responsible Organization: AMSAT-UK/AMSAT-NL
LUSat-OSCAR 74
    Launched: 11/21/2013
    Responsible Organization: Argentinian Ministry of Science, Technology and Productive Innovation
Louisiana-OSCAR 75
    Launched: 11/20/2013
    Responsible Organization: University of Louisiana at Lafayette
Morehead-OSCAR 76
    Launched: 11/21/2013
    Responsible Organization: Morehead State University
CubeSat-OSCAR 77
    Launched: 2/27/2014
    Responsible Organization: University of Tokyo/Tama Art University
LituanicaSAT-OSCAR 78
    Launched: 1/9/2014
    Responsible Organization: Vilnius University

European-OSCAR 79
    Launched: 6/19/2014
    Responsible Organization: The von Karman Institute for Fluid Dynamics
European-OSCAR 80
    Launched: 6/19/2014
    Responsible Organization: The von Karman Institute for Fluid Dynamics
Fuji-OSCAR 81
    Launched: 12/3/2014
    Responsible Organization: Tama Art University
Fuji-OSCAR 82
    Launched: 12/3/2014
    Responsible Organization: Kagoshima University
Navy-OSCAR 83
    Launched: 5/20/2015
    Responsible Organization: US Naval Academy
Navy-OSCAR 84
    Launched: 5/20/2015
    Responsible Organization: US Naval Academy
AMSAT-OSCAR 85
    Launched: 10/8/2015
    Responsible Organization: AMSAT
Indonesia-OSCAR 86
    Launched: 9/28/2015
    Responsible Organization: LAPAN
LUSEX-OSCAR 87
    Launched: 5/30/2016
    Responsible Organization: Satellogic S.A./AMSAT Argentina
Emirates-OSCAR 88
    Launched: 2/15/2017
    Responsible Organization: Mohammed bin Rashid Space Centre/ American University of Sharjah
Tsukuba-OSCAR 89
    Launched: 1/16/2017
    Responsible Organization: University of Tsukuba Satellite Project
LilacSat-OSCAR 90
    Launched: 5/25/2017
    Responsible Organization: Harbin Institute of Technology
AMSAT-OSCAR 91
    Launched: 11/18/2017
    Responsible Organization: AMSAT
AMSAT-OSCAR 92
    Launched: 1/12/2018
    Responsible Organization: AMSAT
Lunar-OSCAR 93
    Launched: 5/18/2018
    Responsible Organization: Harbin Institute of Technology

Lunar-OSCAR 94
    Launched: 5/18/2018
    Responsible Organization: Harbin Institute of Technology

AMSAT-OSCAR 95
    Launched: 12/3/2018
    Responsible Organization: AMSAT

VUSat-OSCAR 96
    Launched: 12/3/2018
    Responsible Organization: Exseed Space

Jordan-OSCAR 97
    Launched: 12/3/2018
    Responsible Organization: Crown Prince Foundation of Jordan

Fuji-OSCAR 98
    Launched: 1/18/2019
    Responsible Organization: Tokyo Institute of Technology

Fuji-OSCAR 99
    Launched: 1/18/2019
    Responsible Organization: Nihon University College of Science and Technology/Japan Amateur Satellite Association (JAMSAT)

Qatar-OSCAR 100
    Launched: 11/15/2018 – the first amateur radio transponder in geostationary orbit.
    Responsible Organization: AMSAT Deutschland e.V./Qatar Amateur Radio Society/Es'hailSat

Philippines-OSCAR 101
    Launched: 10/29/2018
    Responsible Organization: PHL-Microsat/Tohoku University/Hokkaido University

BIT Progress-OSCAR 102
    Launched: 7/25/2019
    Responsible Organization: CAMSAT/Beijing Institute of Technology

Navy-OSCAR 103
    Launched: 6/25/2019
    Responsible Organization: US Naval Academy

Navy-OSCAR 104
    Launched: 6/25/2019
    Responsible Organization: US Naval Academy
    (Information courtesy of AMSAT-NA)

When this book was published in 2020, several amateur radio-capable satellites were reaching orbit each year. At the same time, older satellites have gone silent due to equipment failures, or simply because they re-entered Earth's atmosphere and were destroyed.

The good news is that satellite technology continues to be one of the most healthy and vibrant activities in amateur radio today. The bad news is that changes in the active satellite "fleet" are occurring at rates that make it impossible for any printed book to provide current information about the status of all spacecraft — new or old.

With that in mind, this book will concentrate on the aspects of satellite operating that rarely change: Understanding satellite orbits, learning how to know when a satellite will be available in your area, and understanding how to make contacts through the satellites that interest you, including the kind of hardware and software you'll need to do so.

Your best resource for keeping up to date with satellites that are currently operational, as well as satellites about to be launched, is AMSAT. They have an excellent website at **www.amsat.org** where you can find the status and operating frequencies of all active amateur radio satellites, along with several helpful tools.

Chapter 2
Mysteries of Where and When

One of the many things that sets satellite operating apart from other amateur radio activities is the fact that you are trying to communicate through a moving object. Not only is the spacecraft moving at thousands of miles per hour, the planet you're standing on is turning at a pretty good clip, too!

You don't have to be a proverbial rocket scientist to realize that this could create a complicated situation. You're moving, the satellite is moving, and somehow you must predict when both of you will be in the same place at the same time. Thanks to modern software, making these predictions is relatively easy. There are just a few things you need to know first.

Where Are *You*?

Most people are familiar with the concepts of *latitude* and *longitude*. I could devote this entire chapter to spherical geometry, but it would be a grand waste of time. When it comes to enjoying amateur satellites, just a bedrock understanding will do.

Let's state things simply. Lines of latitude circle the globe horizontally. The equator is at zero degrees latitude and the remaining lines run in parallel with the equator, both north and south, and are labeled by degrees. The degrees measure the angle created by an imaginary line that extends from the center of the Earth to the equator and another line that reaches from the center of the Earth to your location on the globe.

Latitudes north of the equator are expressed with positive numbers; latitudes south of the equator are expressed with negative numbers. The city where I live, for example, is close to +42 degrees north latitude. Buenos Aires, Argentina, way to the south of me (and the equator), is at about -36 degrees south latitude.

Lines of longitude, in contrast, circle the planet vertically. The line that runs through Greenwich, England is zero degrees longitude. It is known as the *Prime Meridian.*

If your location is west of Greenwich, you are on one of the western lines of longitude; if you happen to be east of Greenwich, you are standing on one of the eastern lines of longitude. As with latitude, longitude is measured in degrees.

When you put these two concepts together, it doesn't take much effort to conclude that you can label any point in the world according to its latitude and longitude. By breaking the degrees into "minutes" and "seconds," you can specify the position to an extreme level of precision.

Fortunately, you do not need to determine your location with extreme precision to make a contact through a satellite. When it comes to your location, "close enough" is good enough. You can simply use the latitude and longitude of a city near your location, even if that city is 100 miles away. The easiest way to find this information is on a website known as LatLong (**www.latlong.net**). Just type in the name of the city, followed by the country, and you'll have the coordinates within seconds.

The Maidenhead Locator System

In the early 1980s, a group of amateurs attempted to devise a simple system for determining one's location to a reasonable degree of certainty — not as precise as latitude and longitude coordinates, but close enough for hobby use. The result was the *Maidenhead Locator System* (Maidenhead was the city in

The VHF/UHF Century Club with Satellite Endorsement, better known as VUCC Satellite, is the most sought-after award in the amateur satellite community. To earn your initial VUCC certificate, you must contact at least one station in 100 different grid squares. That's why you'll often be asked for your grid square designator when you make a contact.

England where the group met to finalize the system).

The system divides the world into 324 squares, each one defined as being 1° of latitude by 2° of longitude. Each square, or *grid square* as they are commonly called, has a designation that can include as many as six characters, but for most amateur radio applications we use only four.

A four-character designation is comprised of two letters and two digits. For instance, my home is in grid square FN31.

But if latitude and longitude coordinates can provide a precise description of any location on the planet, why do you need to care about grid squares?

The answer has to do with awards. There are awards available for contacting stations in various grid squares and the ARRL VHF/UHF Century Club Satellite Award is the granddaddy of them all. Hams throughout the world passionately pursue the VUCC, as it is called, so when they contact you through a satellite, they will want to know your grid square designation. If you are chasing awards, you'll also want to know theirs.

Contacts through the FM repeater satellites, which we'll discuss later in this book, tend to be brief exchanges of grid information. For example:

"N6ATQ Echo Mike 87." (Using the phonetic alphabet, he means grid square EM87.)
"Thanks! I'm WB8IMY in Fox Nancy 31."

The easiest way to determine your grid square is to go on the web to a site such as **www.levinecentral.com/ham/grid_square.php**. Just plug in your postal zip code and the site will return your six-digit grid square designation. Remember that you only need the first four characters, reading from left to right.

Where is the *Satellite*?

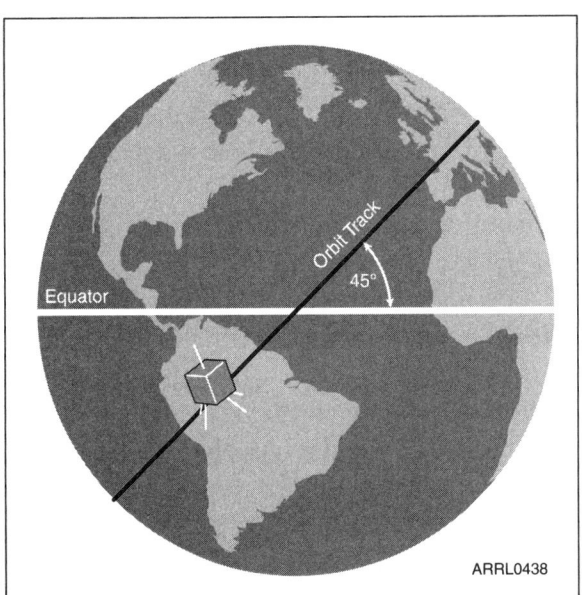

Figure 2.1 — An inclined orbit is one that is inclined with respect to the Earth's equator. In this example, the satellite's orbit is inclined at 45° to the equator.

Now we get to the fun part! Satellites are called *birds* in technical slang for a very good reason: they are constantly flying about, appearing in one part of the sky, zooming overhead, and then quickly disappearing over the horizon. The trick to "catching" a bird is knowing when it is coming and what its flight path may be.

Types of Orbits

Let's take a moment to discuss the paths in which these birds travel. It will help you to understand how predictions are made with your chosen software.

Most amateur satellites are in various types of *Low Earth Orbits* (LEOs), although there are satellites planned for future launch that will travel in the *High Earth Orbits* (HEOs) and one satellite, QO-100, that is in a *Geostationary Orbit* over the Middle East. Let's take a brief look at several of the most common orbits.

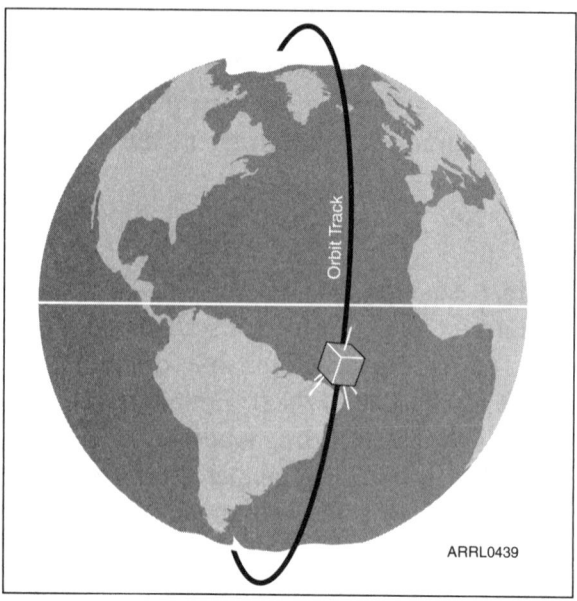

Figure 2.2 — A *sun-synchronous* orbit takes the satellite over the north and south poles. A satellite in this orbit allows every station in the world to enjoy at least one high-elevation pass per day.

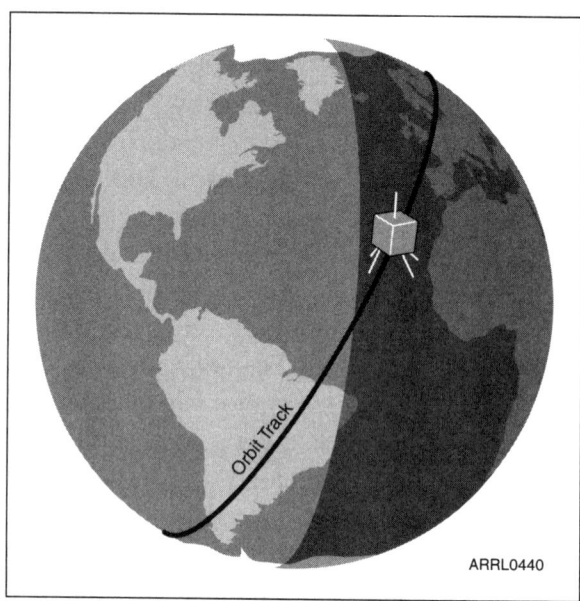

Figure 2.3 — A *dawn-to-dusk* orbit is a variation on the sun-synchronous model except that the satellite spends most of its time in sunlight and relatively little time in eclipse.

An *inclined* orbit is one that is inclined with respect to the Earth's equator. See **Figure 2.1**. A satellite that is inclined 90° would be orbiting from pole to pole; smaller inclination angles mean that the satellite is spending more time at lower latitudes. The International Space Station (ISS), for example, travels in an orbit that is inclined about 50° to the equator. Satellites that move in these orbits frequently fall into the Earth's shadow (eclipse), so they must rely on battery systems to provide power when the solar panels are not illuminated. Depending on the inclination angle, some locations on the Earth will never have good access because the satellites will rarely rise above their local horizons. This was true, for example, in the days when the US Space Shuttles carried amateur radio operators. Shuttle orbits were usually inclined at low angles and hams living in the northern US rarely enjoyed passes that brought the Shuttle to a decent elevation above their local horizon.

A *sun-synchronous* orbit takes the satellite over the north and south poles. See **Figure 2.2**. There are two advantages to a sun-synchronous orbit: (1) the satellite is available at approximately the same time of day, every day, and (2) everyone, no matter where they are, will enjoy at least one high-altitude pass per day.

A *dawn-to-dusk* orbit is a variation on the sun-synchronous model except that the satellite spends most of its time in sunlight and relatively little time in eclipse. See **Figure 2.3**.

The *Molniya* High Earth Orbit (**Figure 2.4**) was pioneered by the former Soviet Union. It is an elliptical orbit that carries the satellite far into space at its greatest distance from Earth (apogee). To observers on the ground, the satellite at apogee appears to hover for hours at a time before it plunges earthward and sweeps to (sometimes) within 1,000 kilometers or so of the Earth at its closest approach (perigee). One great advantage of the Molniya orbit is that the satellite

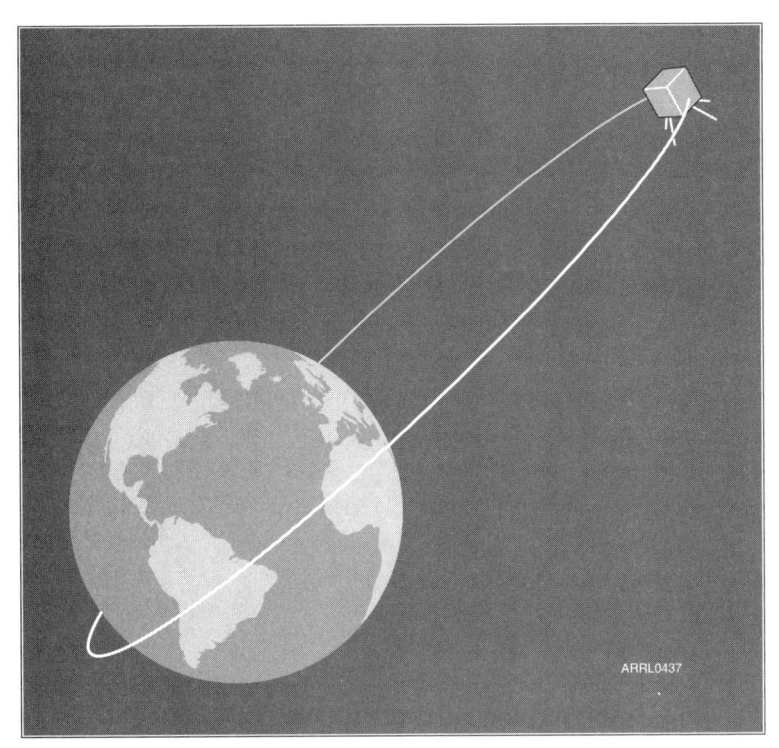

Figure 2.4 — The *Molniya* orbit is an elliptical orbit that carries the satellite far into space at its greatest distance from Earth (apogee). To observers on the ground, the satellite at apogee appears to hover for hours at a time before it plunges earthward and (often) sweeps within 1,000 kilometers at its closest approach (perigee).

is capable of "seeing" an entire hemisphere of the planet while at apogee. Hams can use a Molniya satellite to enjoy long, leisurely conversations spanning thousands of kilometers here on Earth. When this book was being written, there were no active Molniya hamsats in orbit.

In a geostationary orbit, the satellite is positioned about 22,000 miles from Earth and its speed matches the rotational speed of our planet at the equator. The result is a satellite that appears to be fixed at one point in the sky, at least from our perspective here on Earth. Geostationary orbits are ideal for communications because the satellite offers extremely wide coverage and is available around the clock. Your antenna doesn't need to track the satellite; you just point it in the direction of the bird, and you're done.

Attractive as they may be for communications, geostationary orbits are difficult to reach, and they require highly expensive launch operations to get there. Moreover, once a satellite reaches its assigned position, it must be regularly monitored, and its position maintained. That's why the only amateur radio transponder aboard a geostationary satellite — QO-100 — is part of a much larger commercial spacecraft.

Satellite Footprints

Speaking of how much of our planet a satellite sees, it is important to understand the concept of the satellite's *footprint*. A satellite footprint can be loosely defined as the area on the Earth's surface that is "illuminated" by the satellite's antenna systems at any given time. Another way to think of a footprint is to regard it as the zone within which stations can communicate with each other through the satellite.

Unless the satellite in question is geostationary, footprints are constantly moving. Their sizes can vary considerably, depending on the altitude of the satellite. The footprint of the low-orbiting International Space Station is about 600 kilometers in diameter. In contrast, a higher orbiting satellite may have a footprint of more than 1,500 kilometers across. See the example of a satellite footprint in **Figure 2.5**. The amount of time you have available to communicate depends on how long your station remains within the footprint. This time can be measured in minutes, or in the case of a satellite in a Molniya orbit, hours.

It is worthwhile to note that the size and even the shape of a footprint can

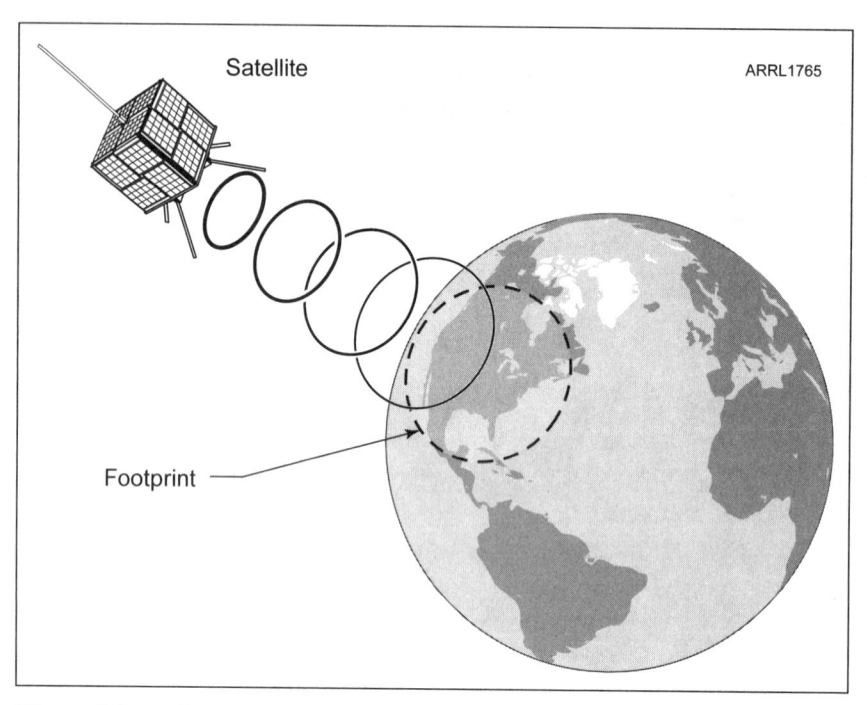

Figure 2.5 — A satellite's footprint is the area within which two stations can theoretically communicate.

also vary according to the type of antenna the satellite is using. A highly directional antenna with a narrow beamwidth will create a small footprint even though the satellite is traveling in a high-altitude orbit. This usually isn't an issue for amateur satellites, however.

The Satellite's Position Relative to You

You know a satellite is coming your way because your software has told you so. You are standing in your backyard, holding a portable antenna, and waiting for the action to start. This is all well and good, but you need to know the satellite's path across your local sky so that you can point your antenna in the proper direction.

Fortunately, your software has provided this valuable information in terms of the satellite's *azimuth* and *elevation* relative to your station.

Azimuth describes the satellite's position in degrees referenced to true north. See **Figure 2.6**, and imagine yourself in the center of a giant compass circle that is divided in degree increments from 0 to 360. North is 0° (it is also 360°), east is 90°, south is 180°, and west is 270°. If your tracking software indicates that you need to point your antenna to an azimuth of 135° to intercept the satellite, for example, you're going to point the antenna southeast.

Let's take a look at a more detailed example. Once again, your station in **Figure 2.7** is in the center of the compass circle. According to your satellite tracking program, the International Space Station (ISS) is scheduled to rise above your local horizon at precisely 03:57:30 UTC. The program may describe the satellite's azimuth path like this:

Time	Azimuth (degrees)
03:57	307
03:58	350
03:59	0
04:00	11
04:01	20
04:02	30

When you plot these azimuth points, you can quickly see the horizontal path the satellite is going to take. The bird is going to rise in your northwestern sky and quickly move toward the east, finally dipping below your northeast horizon at about 30°. If you have rotating antennas, you can see that they'll need to be pointing northwest at the beginning of the satellite's pass. As the satellite moves across the sky, your antenna will need to track around the circle from 307°, to 0°, and so on until they are pointing at 30° azimuth when the satellite finally disappears.

Let's add another dimension to our satellite track — *elevation*. Elevation is simply the angle, in degrees, between your station and the satellite, referenced to the Earth's surface. See **Figure 2.8**. The elevation angle begins at 0° with the satellite at the horizon and increases to 90° when the satellite is directly overhead. Elevation is every bit as critical as azimuth if you are using directional antennas. Not only do your antennas need to be pointed at the satellite as it appears to move in the horizontal plane,

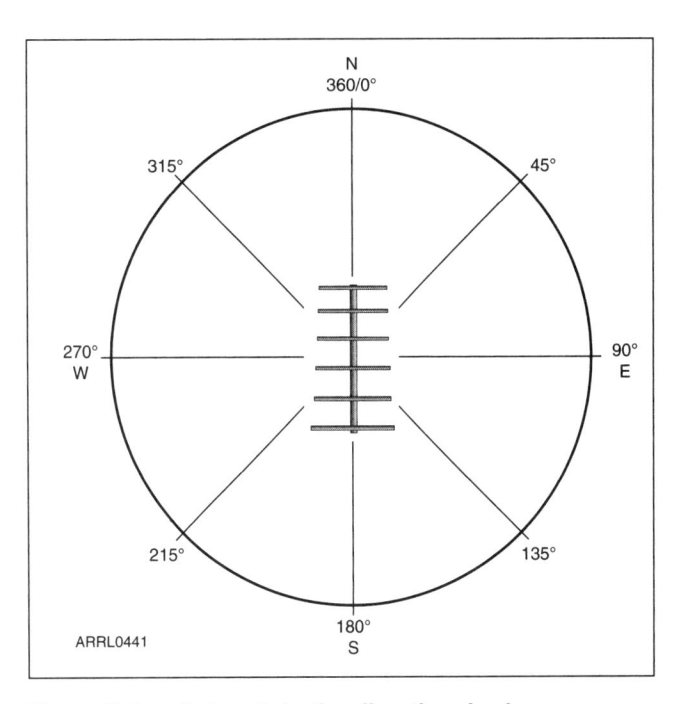

Figure 2.6 — Azimuth is the direction, in degrees referenced to true north, that an antenna must be pointed to receive a satellite signal. Imagine your station in the center of a giant compass circle that is divided in degree increments from 0 to 360. North is 0° (actually, it is also 360°), east is 90°, south is 180°, and west is 270°.

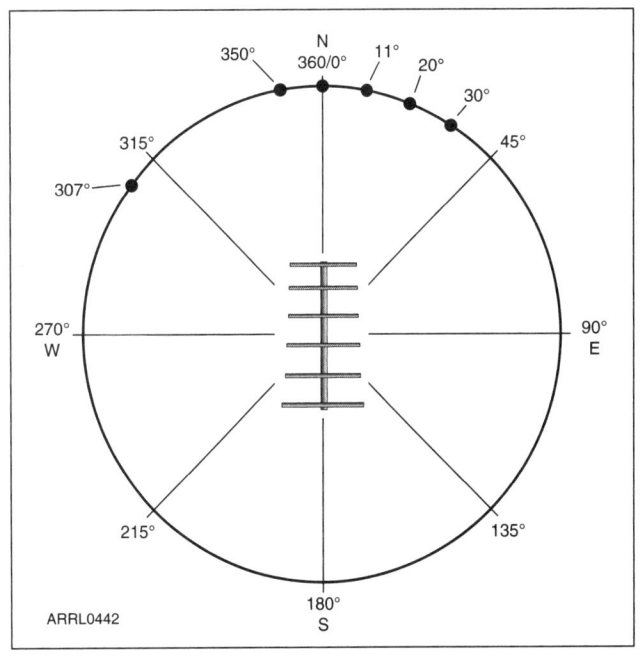

Figure 2.7 — The azimuth path of the ISS for our hypothetical pass.

Mysteries of Where and When 2-7

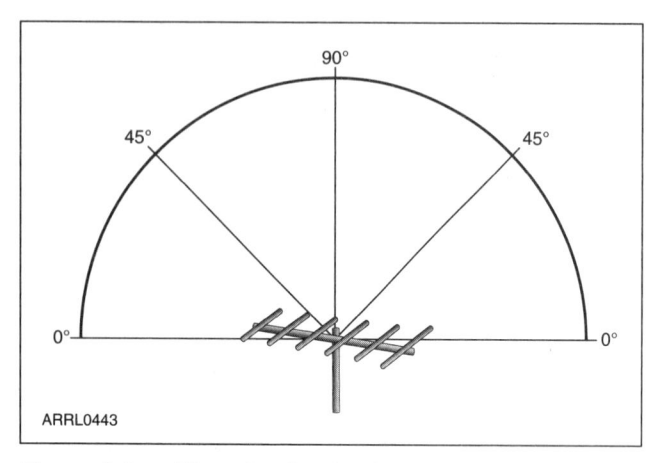

Figure 2.8 — Elevation is simply the angle, in degrees, between your station and the satellite, referenced to the Earth's surface.

they must also tilt up and down to track the satellite as it moves in the vertical plane.

Many amateur satellite stations use devices known as *az/el* (azimuth/elevation) *rotators* to move their directional antennas in both planes as the satellite streaks across the sky. However, az/el rotators are not strictly necessary to enjoy satellite operating. We'll discuss this in detail later.

Even if you are not using movable antennas, knowing a satellite's elevation track is important for another reason. Unless you live in Kansas or a similarly flat location, chances are you do not have a clear view to the horizon in every direction. Perhaps there are serious RF obstacles such as mountains, hills, or buildings blocking the way. If you are trying to receive a microwave signal from a satellite, the RF absorption properties of trees can present serious obstacles, too. The elevations of these objects represent your true *radio horizons* in whichever direction they may lie (**Figure 2.9**). If you have a ridge to the north with a maximum elevation of 30° above the horizon as viewed from your station, your northern radio horizon *begins* at 30° elevation. You can't communicate with a satellite in your northern sky until it rises above 30°, so you'll have to take

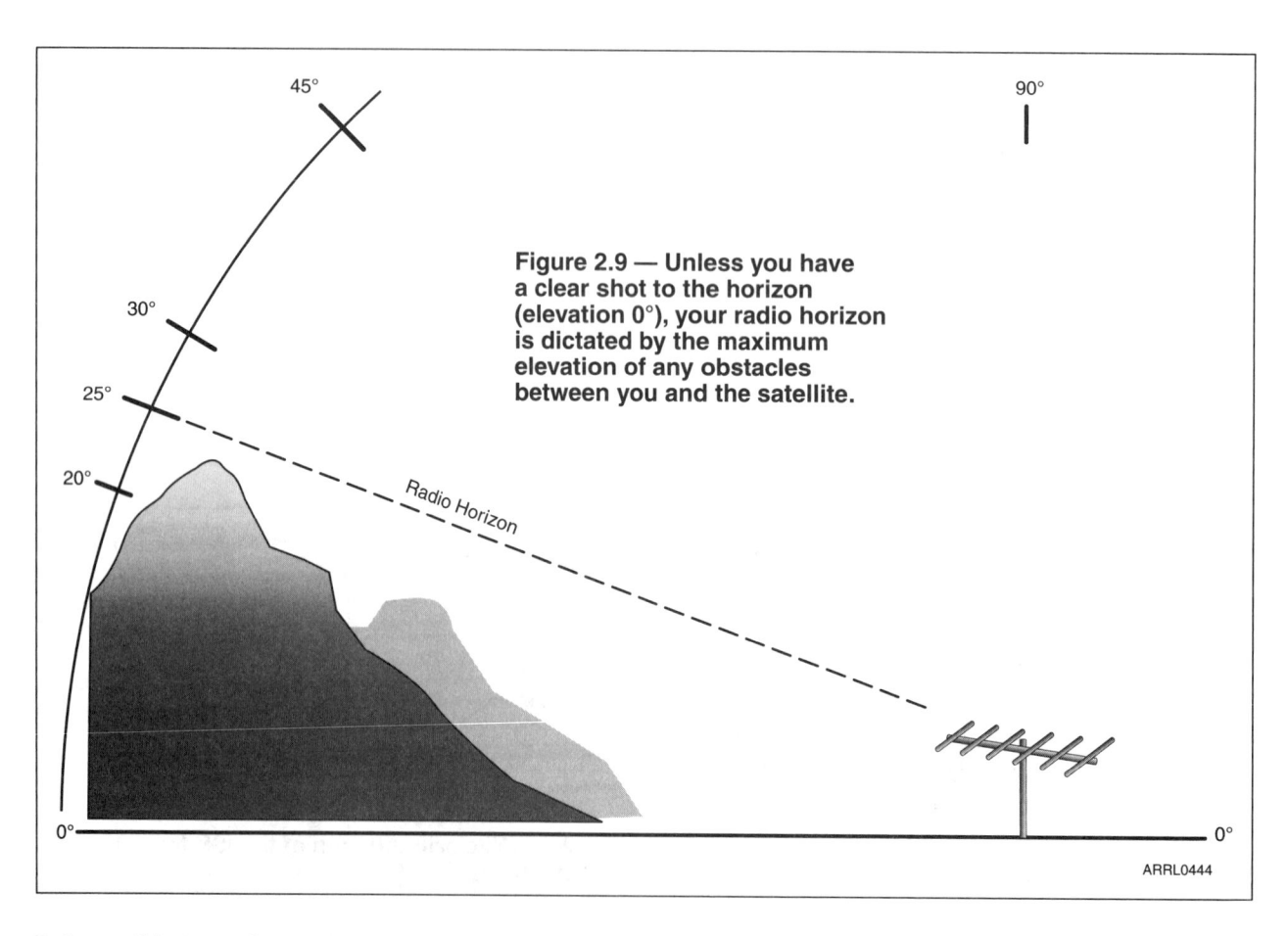

Figure 2.9 — Unless you have a clear shot to the horizon (elevation 0°), your radio horizon is dictated by the maximum elevation of any obstacles between you and the satellite.

that fact into account when you view the information provided by your satellite tracking software.

Usually, and particularly for satellites in low Earth orbits, as the satellite's elevation angle increases, its distance from you decreases. This is a good thing since the closer the satellite, the stronger the radio signal. With that idea in mind, the higher the elevation of a satellite pass, the better, right? Well…yes and no. Remember that satellites are moving at high speeds relative to your position. As they move closer to you (move higher in elevation), the *Doppler Effect* increasingly comes into play.

The Doppler Effect

The Doppler Effect, named after scientist Christian Doppler (1803 – 1853), is the apparent change in frequency of sound or electromagnetic waves, varying with the relative velocity of the source and the observer. See **Figure 2.10**. Thanks to the Doppler Effect, as a satellite moves toward your location, its signal will *increase* in frequency; as it moves away from you, its signal will *decrease* in frequency.

When considering the Doppler Effect, it is important to realize that the satellite's transmit frequency is *not* changing. What is changing is the frequency of its signal *at your station*. To understand this, consider the following baseball analogy illustrated in **Figure 2.11**. A baseball pitcher throws one ball every second and the ball takes one second to travel the distance between the pitcher's mound and home plate. As long as the pitcher and catcher remain fixed in their relative positions, nothing will change. This is exactly the condition present when a satellite is geostationary.

However, if the pitcher begins moving toward the catcher, the time it takes for the ball to travel between the pitcher and the catcher begins to *decrease*. From the pitcher's point of view, nothing has changed. He is still tossing balls at a rate of one per second. From the catcher's point of view, however, there has been a noticeable change. Before the pitcher began moving, the balls were smacking into the catcher's glove exactly one second after they left the pitcher's hand. Now the balls are arriving in less than one second. As the pitcher continues to move closer, the balls arrive sooner, one after the other. The pitcher is still throwing balls at a rate of one per second, but the time interval between the throw and the catch is shrinking *at a rate proportional to the speed at which the pitcher is approaching the catcher*.

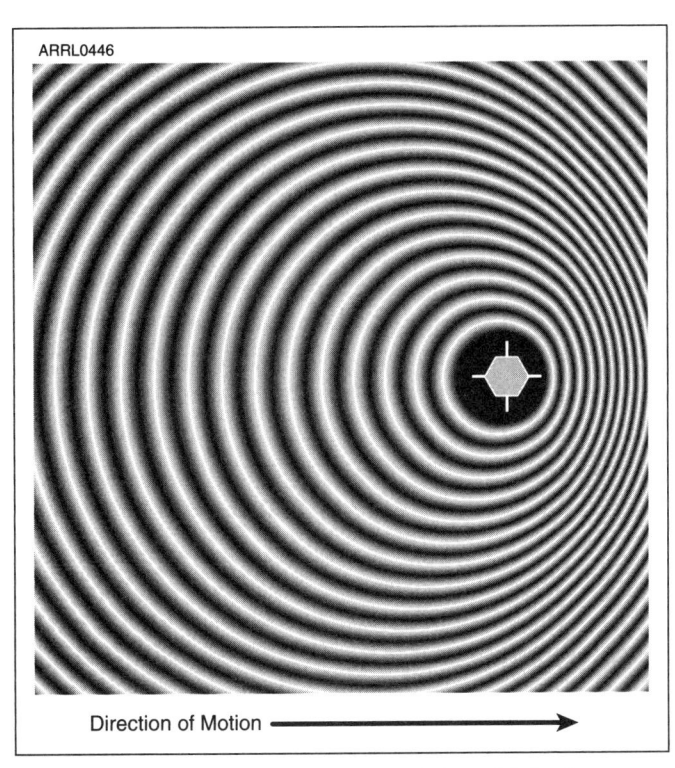

Figure 2.10 — It is sometimes helpful to think of the Doppler Effect as being caused by radio waves "crowding up" as a satellite moves toward your location. As a result, its signal will appear to *increase* in frequency as it moves toward you; as it moves away from you, its signal will *decrease* in frequency.

Mysteries of Where and When 2-9

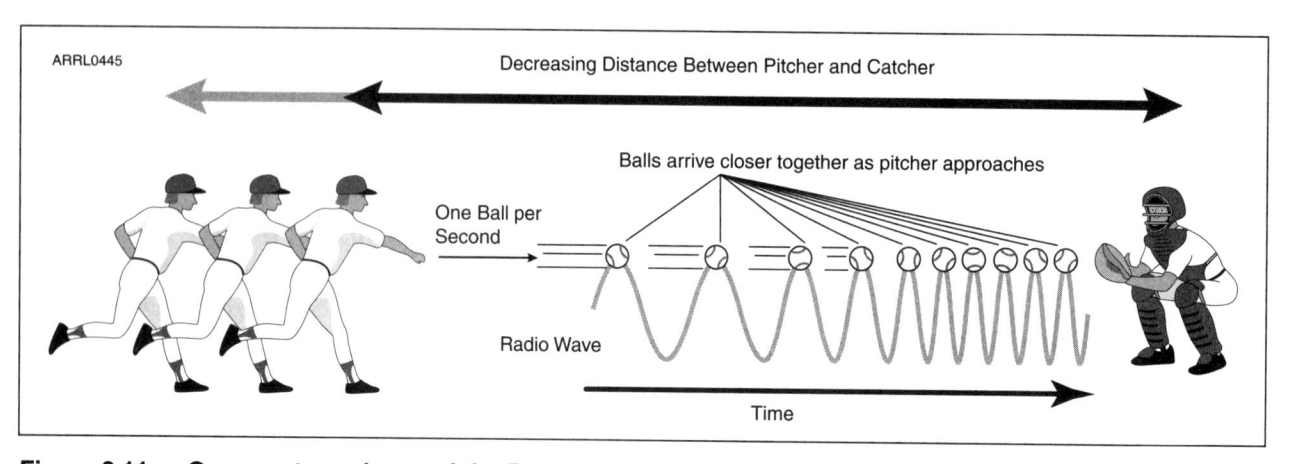

Figure 2.11 — One way to understand the Doppler Effect is by imagining a baseball pitcher who throws one ball every second. The ball takes one second to travel the distance between the pitcher's mound and home plate. If the pitcher is stationary, the catcher will receive one ball every second. That's because the velocity of the ball and the distance between the pitcher and the catcher remain unchanged. However, if the pitcher begins moving toward the catcher, the time it takes the ball to travel between the pitcher and the catcher decreases. If you imagine each ball representing the crest of a wave and the wavelength being the distance between one ball and another, the wavelength is decreasing as the pitcher moves toward the catcher. Since a shorter wavelength translates to a higher frequency, the frequency appears to increase. Conversely, as the wavelength increases, the frequency decreases.

If you imagine each ball representing the crest of a wave and the wavelength being the distance between one ball and the next, the wavelength is decreasing as the pitcher moves toward the catcher. Thinking in terms of an RF signal, a shorter wavelength translates to a higher frequency. Substitute your satellite station for the catcher, and you can see how this Doppler Effect results in a higher received frequency as the "pitcher" (satellite) moves closer. Conversely, as the satellite moves away, the wavelength increases and the received frequency decreases.

You probably experience the Doppler Effect almost every day. When a fire truck approaches at high speed on a nearby freeway, you hear its siren blaring at a higher pitch, shifting downward as the truck passes and speeds off into the distance. The same thing happens with satellites, but unlike an earthbound fire truck that is moving at 60 MPH, the satellite is screaming by at thousands of miles per hour. The proportional difference between your speed and the speed of the satellite is enormous — high enough to shift the received frequency of a radio wave!

On a practical level, a high-elevation satellite pass can be problematic because the frequency shift caused by the Doppler Effect can be considerable. The effect also increases the higher you move in frequency. It can be quite a juggling act to adjust your receiver while trying to carry on a conversation. We'll discuss the operational aspects of coping with Doppler in a later chapter. For now, suffice to say that while high elevation passes are best for signal strength, they present their own challenges thanks to the Doppler Effect.

Azimuth and Elevation Combined

Let's combine azimuth and elevation for a truly realistic satellite track, using our previous example of the International Space Station (ISS). We'll add the station's downlink frequency so we can see the Doppler Effect in action.

Time	Azimuth (degrees)	Elevation (degrees)	Frequency (MHz)
03:57	307	0	145.804
03:58	350	10	145.803
03:59	0	18	145.800
04:00	11	9	145.798
04:01	20	5	145.797
04:02	30	0	145.795

In this example, the ISS rises to an elevation of 18° at 03:59 UTC before sinking back down to the horizon at 04:02 UTC. This is considered a low-elevation pass. If you have objects in your northern sky that rise above 18° elevation, you won't be able to communicate with the space station during this pass. The space station is transmitting at 145.800 MHz, but you'll notice that the frequency change caused by the Doppler Effect is minimal because the distance and relative velocity between you and the space station doesn't change dramatically.

Now we'll modify our example, making it a high-elevation pass.

Time	Azimuth (degrees)	Elevation (degrees)	Frequency (MHz)
03:57	307	0	145.810
03:58	350	10	145.808
03:59	0	25	145.806
04:00	11	40	145.804
04:01	20	65	145.802
04:02	30	80	145.800
04:03	36	60	145.798
04:04	41	45	145.796
04:05	50	29	145.794
04:06	55	15	145.792
04:07	59	0	145.790

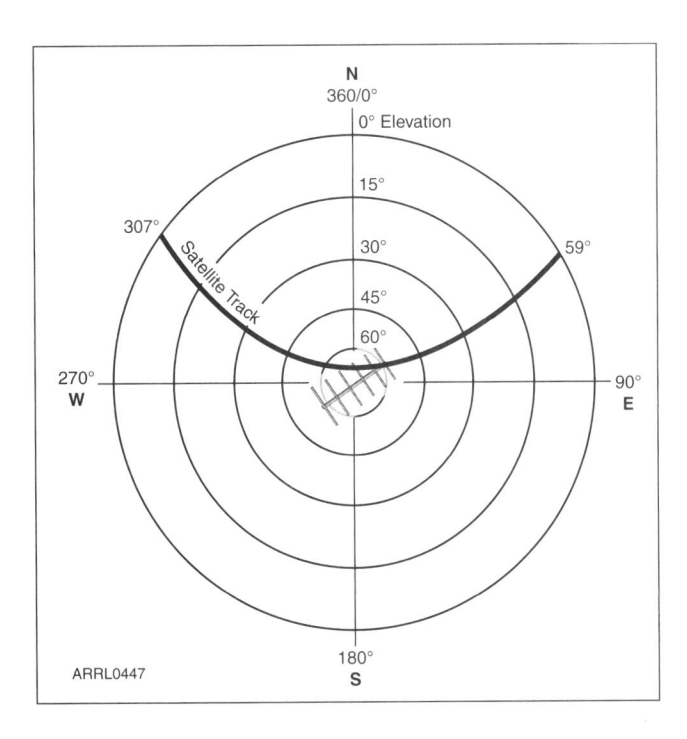

This pass of the ISS is also plotted graphically in **Figure 2.12**. In this illustration, we combine the azimuth and elevation, creating a "radar screen" display with your station in the center.

There are several interesting things to note in this example. Did you notice that this high-elevation pass (topping out at 80° at 04:02 UTC) had a longer overall duration than the pervious low-elevation pass? The low-elevation pass lasted only 5 minutes; this pass was a full 10 minutes in length. Obviously, when an object is tracking to a high elevation

Figure 2.12 — In this illustration, we combine the azimuth and elevation, creating a "radar screen" display of a satellite pass with your station in the center. The concentric rings represent elevation; the outer ring is azimuth.

Mysteries of Where and When 2-11

in the sky (almost directly overhead in this example), it is in the sky for a longer period.

Did you also notice what the Doppler Effect did to the downlink signal frequency at your station? Because the distance and relative velocity between you and the space station changed substantially during the pass, the Doppler Effect was very much in play. The result was a receive frequency that began at 145.810 MHz, shifted down to 145.800 MHz at maximum elevation, and then continued downward until it reached 145.790 MHz as the station slipped below the horizon. That's a 20 kHz frequency shift throughout the pass!

Satellite Tracking Applications

Satellite operators get their pass predictions using one of three sources:
- Software running on laptops or desktop computers.
- Applications (apps) on smartphones or tablets.
- Satellite tracking websites.

Software for Laptops and Desktops

You'll find satellite software programs written for Windows, Mac, and Linux operating systems. Several popular applications are listed in **Table 2.1**. When computers were first employed to track amateur satellites, they provided only the most basic, essential information: when the satellite will be available (AOS, acquisition of signal), how high the satellite will rise in the sky, and when the satellite is due to set below your horizon (LOS, loss of signal). Today we tend to ask a great deal more of our tracking programs. Modern applications still provide the basic information, but they usually offer more features such as:
- Predicted frequency offset (Doppler shift) on the link frequencies.
- The orientation of the spacecraft's antennas with respect to your ground station and the distance between your ground station and the satellite.
- Which regions of the Earth have access to the spacecraft; that is, who's in range?
- Whether the satellite is in sunlight or being eclipsed by the Earth. Some spacecraft only operate when in sunlight.
- When the next opportunity to cover a selected terrestrial path (mutual window) will occur.

Several applications do even more. Some will control antenna rotators,

Table 2.1
Satellite Tracking Software for Desktops and Laptops

Name	Source	Operating System
SatPC32	www.amsat.org/product-category/software/	Windows
SatScape	scotthather.weebly.com/satscape.html	Windows, MacOS, Linux
MacDoppler	www.dogparksoftware.com/MacDoppler.html	MacOS
Predict	www.qsl.net/kd2bd/predict.html	Linux
Gpredict	gpredict.oz9aec.net/	Linux

automatically keeping directional antennas aimed at the target satellite. Other applications will also control the radio to automatically compensate for frequency changes caused by Doppler shifting.

PC/Laptop Software Tips

There are so many different types of satellite software, and they change so frequently, it would be foolhardy to attempt to give you detailed operational descriptions in any book.

Even so, there are a number of aspects of satellite-tracking software that rarely change. For example, we spent some time discussing how to determine your location with sufficient accuracy to be useful for satellite tracking. The next step is to get that information into your chosen program.

Most programs will ask you to enter your station location as part of the initial setup process. Some applications use the term "observer" to mean "station location," but the terms are synonymous for the sake of our discussion. Sophisticated programs will go as far as to provide you with a list of cities that you can select to quickly enter your location. Other programs will ask you to enter your latitude and longitude coordinates manually.

When entering latitude, longitude, and other angles, make sure you know whether the computer expects degree-minute or decimal-degree notation. Following the notation used by the on-screen prompt usually works. Also make sure you understand the units and sign conventions being used. For example, longitudes may be specified in negative numbers for locations west of Greenwich (0° longitude). Latitudes in the southern hemisphere may also require a minus sign. Fractional parts of a degree will have very little effect on tracking data, so in most cases you can just ignore it.

Dates can also cause considerable trouble. Does the day or month appear first? Can November be abbreviated as Nov, or must you enter 11? The number

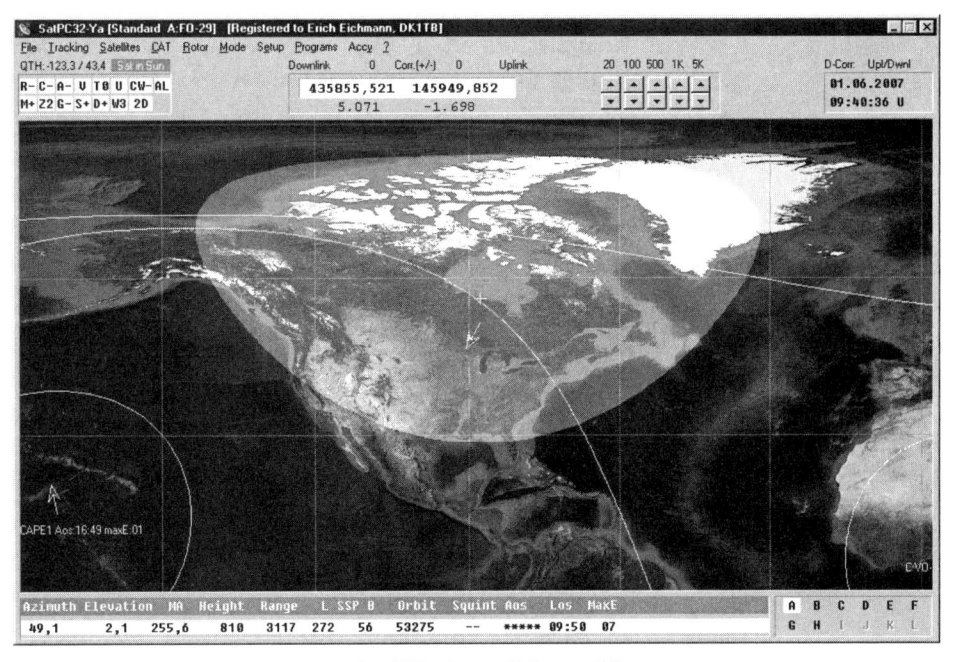

Sat32PC is tracking software for Windows PCs and laptops.

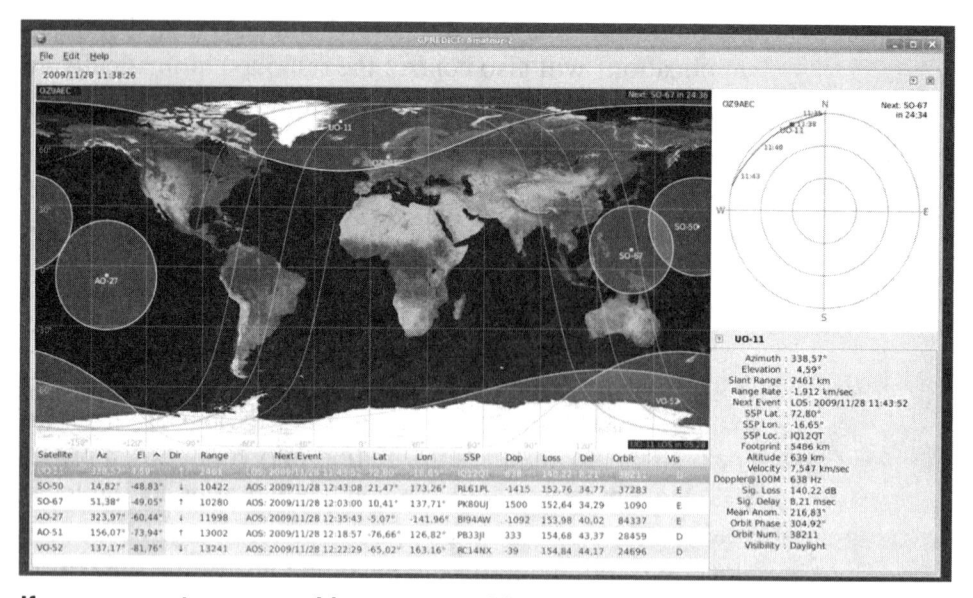

If your computer runs on Linux, one tracking option is *GPredict*.

is almost always required. Must you write 2020 or will 20 suffice? Should the parts be separated by colons, dashes, or slashes? The list goes on and on. Once again, the prompt is your most important clue. For example, if the prompt reads "Enter date (DD:MM:YY)" and you want to enter Feb. 9, 2020, follow the format of the prompt as precisely as possible and write 09:02:20.

Once you have your coordinates entered, you're still not quite done. The software now "knows" its location, but it doesn't know the locations of the satellites you wish to track. The only way the software can calculate the positions of satellites is if it has a recent set of *orbital elements*.

Orbital Elements

Orbital elements are a set of six numbers that completely describe the orbit of a satellite at a specific time. Although scientists may occasionally use different groups of six quantities, radio amateurs nearly always use the six known as Keplerian Orbital Elements, or simply, *Keps*. (Kepler, you may recall, discovered some interesting things about planetary motion back in the 17th century!)

These orbital elements are derived from very precise observations of each satellite's orbital motion. Using precision radar and highly sensitive optical observation techniques, the North American Aerospace Defense Command (NORAD) keeps a very accurate catalog of almost everything in Earth orbit. Periodically, they issue the unclassified portions of this information to the National Aeronautics and Space Administration (NASA) for release to the general public. The information is listed by individual catalog number of each satellite and contains numeric data that describes, in a mathematical way, how NORAD observed the satellite moving around the Earth at a very precise location in space at a very precise moment in the past.

Without getting into the complex details of orbital mechanics (or Kepler's laws) suffice it to say that your software simply uses the orbital element information NASA publishes that describes where a particular satellite was "then" to

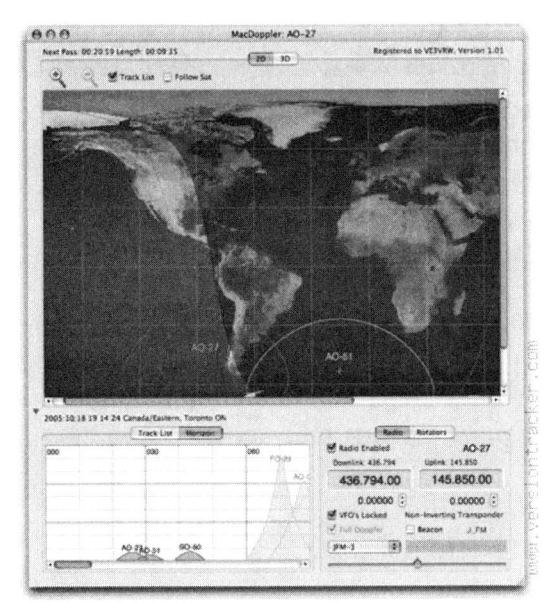

If you own a Mac, you can track amateur satellites with *MacDoppler*.

solve the orbital math and make a prediction (either graphically or in tabular format) of where that satellite ought to be "now." The "now" part of the prediction is based on the local time and station location information you've also been asked to load into your software.

In the old days you had to load orbital elements into your software manually. This is no longer the case. If you have an internet connection, your software will do this automatically. If you are curious to know more about orbital elements, see the sidebar "Understanding Orbital Elements."

Smartphones and Tablets

Although many satellite-active amateurs still use PCs and laptop computers, smartphones and tablet computers are the most popular means of tracking satellites today. This is especially true among hams who enjoy making satellite contacts while camping, hiking, or from vehicles.

It's easy to understand why these devices have come to dominate the amateur satellite world. With a satellite app on your smartphone or tablet, there is no need to know your location; the app will use the Global Positioning System receiver in the device to determine exactly where you are. The app will keep itself up to date with the latest orbital elements whenever it knows an internet

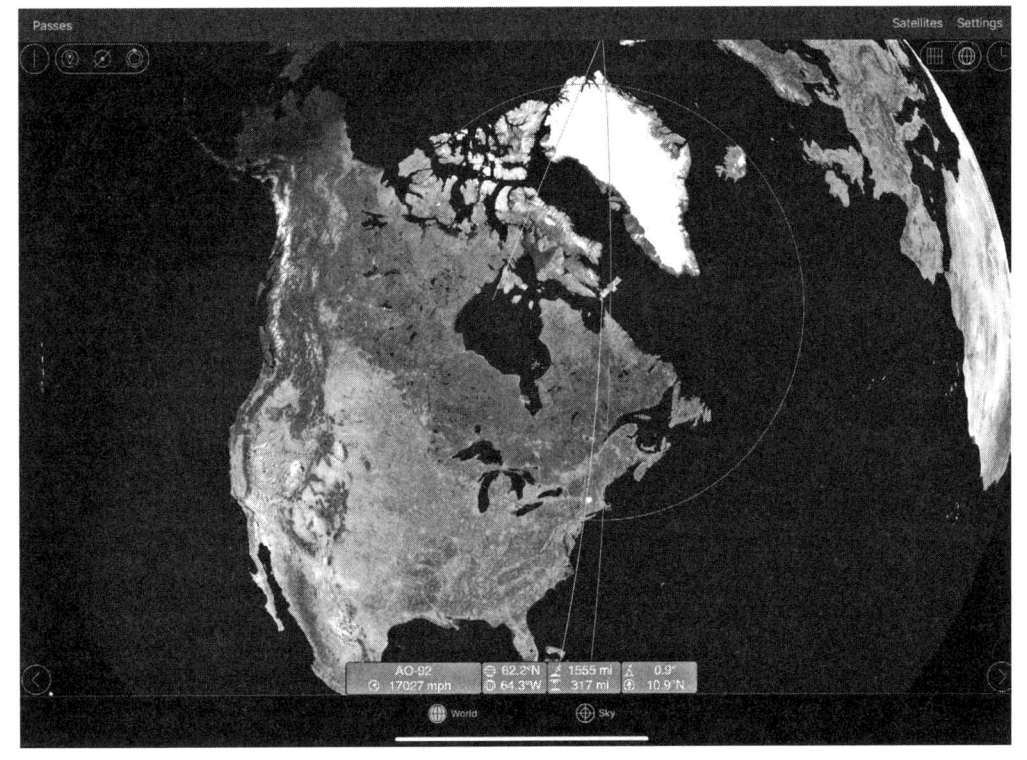

***GoSatWatch* is a popular satellite-tracking app for Apple iPhones or iPads.**

Mysteries of Where and When   2-15

Understanding Orbital Elements

Orbital elements are frequently distributed with additional numerical data (which may or may not be used by a software tracking program) and are commonly available in two forms:

NASA Two-Line Elements, such as the ones shown below for OSCAR 27.

AO-27

1 22825U 93061C 08024.00479406 -.00000064 00000-0 -86594-5 0 8811

2 22825 098.3635 349.6253 0008378 336.4256 023.6532 14.29228459747030

AMSAT Verbose Elements, such as these for OSCAR 51.

Satellite: AO-51
Catalog number: 28375
Epoch time: 08024.16334624
Element set: 14
Inclination: 098.0868 deg
RA of node: 056.7785 deg
Eccentricity: 0.0083024
Arg of perigee: 232.8417 deg
Mean anomaly: 126.5166 deg
Mean motion: 14.40594707 rev/day
Decay rate: 7.0e-08 rev/day^2
Epoch rev: 18752
Checksum: 310

Let's use the easier-to-understand AMSAT format to break down the meaning, line by line.

The first two entries identify the spacecraft. The first line is an informal **satellite name**. The second entry, **Catalog Number**, is a formal ID assigned by NASA.

The next entry, **Epoch Time**, specifies the time the orbital elements were computed. The number consists of two parts, the part to the left of the decimal point that describes the year and day, and the part to the right of the decimal point that describes the (very precise) time of day. For example, 96325.465598 refers to 1996, day 325, time of day .465598.

The next entry, **Element Set**, is a reference used to identify the source of the information. For example, 199 indicates element set number 199 issued by AMSAT. This information is optional.

The next six entries are the six key orbital elements.

Inclination describes the orientation of the satellite's orbital plane with respect to the equatorial plane of the Earth. Recall from earlier in this chapter that the higher a satellite's orbital inclination, the more time the bird spends away from the Equator.

RAAN, Right Ascension of Ascending Node, specifies the orientation of the satellite's orbital plane with respect to fixed stars.

Eccentricity refers to the shape of the orbital ellipse. You may recall our earlier discussion of elliptical Molniya orbits. These orbits are highly eccentric. The value of the eccentricity element also yields some rough information as to the shape of the orbit the satellite is following. The closer this number is to "0", the more circular the orbit of the satellite tends to be. Conversely, an eccentricity value approaching "1," indicates the satellite is following a more elliptically shaped (possibly a Molniya) orbital path. For example, many Molniya orbit satellites have eccentricities in the .6 to .7 range.

Argument of Perigee describes where the perigee of the satellite is in the satellite orbital plane. Recall that a satellite's perigee is its closest approach to the Earth. When the argument of perigee is between 180° and 360° the perigee will be over the Southern Hemisphere. Apogee — a satellite's most distant point from the Earth — will therefore occur above the Northern Hemisphere.

Mean Anomaly locates the satellite in the orbital plane at the epoch. All programs use the astronomical convention for mean anomaly (MA) units. The mean anomaly is 0 at perigee and 180 at apogee. Values between 0 and 180 indicate that the satellite is headed up toward apogee. Values between 180 and 360 indicate that the satellite is headed down toward perigee.

Mean Motion specifies the number of revolutions the satellite makes each day. This element indirectly provides information about the size of the elliptical orbit.

Decay Rate is a parameter used in sophisticated tracking models to consider how the frictional drag produced by the Earth's atmosphere affects a satellite's orbit. It may also be referred to as rate of change of mean motion, first derivative of mean motion, or drag factor. Although decay rate is an important parameter in scientific studies of the Earth's atmosphere and when observing satellites that are about to reenter, it has very little effect on day-to-day tracking of most amateur radio satellites. If your program asks for drag factor, enter the number provided. If the element set does not contain this information enter zero — you shouldn't discern any difference in predictions. You usually have a choice of entering this number using either decimal form or scientific notation. For example, the number -0.00000039 (decimal form) can be entered as $-3.9e-7$ (scientific notation). The $e-7$ stands for 10 to the minus seventh power (or 10 exponent -7). In practical terms $e-7$ just means move the decimal in the preceding number 7 places to the left. If this is confusing, remember that in most situations entering zero will work fine.

Epoch revolution is just another term for the expression "Orbit Number" that we discussed earlier. The number provided here does not affect tracking data, so don't worry if different element sets provide different numbers for the same day and time.

The ***Checksum*** is a number constructed by the data transmitting station and used by the receiving station to check for certain types of transmission errors in data files. It does not bear any relationship to a satellite's orbit.

Table 2.2

Satellite Apps for Smartphones and Tablets

Name	Source	Operating System
Satellite Tracker ProSat	Apple iTunes App Store	iOS
GoSatWatch	Apple iTunes App Store	iOS
Satellite Tracker	Google Play	Android
Heavens-Above	Google Play	Android

connection is available. The app displays are crisp and easy to read, and they provide a wealth of useful information. Many apps will even sound an alert when one of your favorite satellites is about to come streaking by.

Apps are so easy to use, there isn't much to say about them. Most also provide "help" functions to guide you. Of course, the downside of using an app is that it cannot perform some of the fancy tasks such as automatically turning your antennas or adjusting your transceiver frequency to compensate for Doppler shift.

See **Table 2.2** for a list of some of the popular satellite apps that were available when this book was written.

Satellite Tracking on the Web

If you don't own a smartphone or table computer, and don't want to invest in tracking software for a PC, there is an alternative: the AMSAT Online Pass Predictions website at **www.amsat.org/track/**.

On this page you simply select the spacecraft you want, enter your grid square designation, which it uses to calculate your approximate latitude and longitude, and then click the Predict button. The site will produce predictions that you can easily write down or copy and paste into a document that you can print.

Note that the times are UTC, not your local time. You'll have to determine the pass time by subtracting the correct number of hours from UTC. You can find this information online at **www.timetemperature.com/tzus/gmt_united_states.shtml**.

Chapter 3

Your Satellite Station

What constitutes a satellite station? That's a good question, and it has several answers. A satellite station can be extremely simple, or it can be so elaborate that even NASA would be impressed!

You have considerable flexibility, depending on your goals and the amount of money you want to invest. Let's discuss a few options.

Option 1: The Bare Minimum Station

To make contacts through FM repeater satellites, you need only a dual-band (2-meter/70-centimeter) FM transceiver and a dual-band mobile or base-station antenna. There are some serious drawbacks with this setup, however.

A mobile or base-station antenna is, by design, *omnidirectional*. It radiates energy in several directions, but not in a focused manner. In other words, it has little, if any, gain — either in transmitting or receiving. To compensate, you are going to need an FM transceiver with a considerable amount of RF output (about 50 W). Since your antenna has such poor gain, you will need this power to cut through the interference and be heard by the satellite.

Your ability to receive the satellite's signal will also be reduced, so it is likely that you will only be able to make contacts when the bird is at its highest elevation in the sky. You may not even hear the satellite until it is well above the horizon.

All these performance deficits notwithstanding, you can still make many satellite contacts with this type of station. Considering the relatively low prices of new and used dual-band FM transceivers today, this is an economical way to get started.

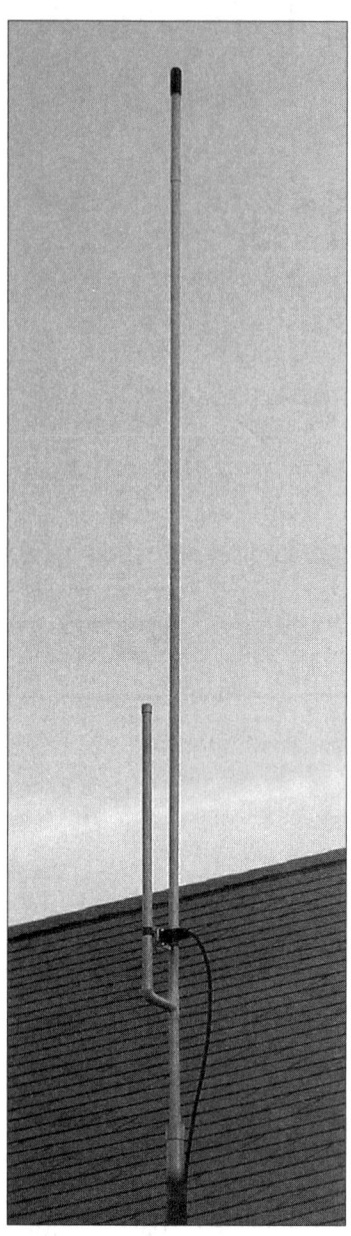

This is an ordinary dual-band J-pole antenna, the kind you would find at many ham stations. Few realize that this same omnidirectional antenna can be used to communicate with FM repeater satellites.

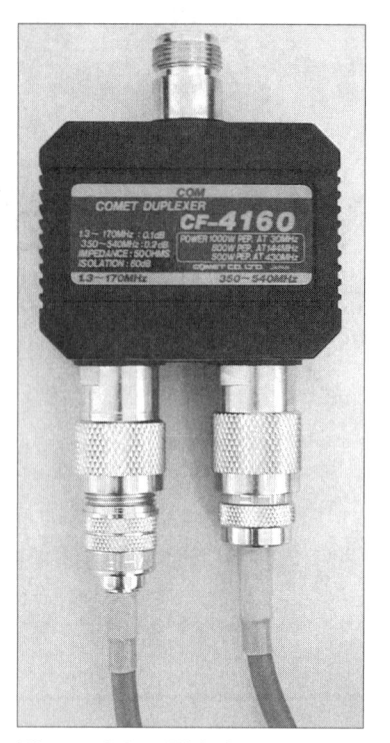

Figure 4.1 — This is a diplexer (often called a duplexer) for 2 meters and 70 centimeters.

Dual-band base and mobile antennas are also commonly available at reasonable prices. One thing to keep in mind, however, is that a dual-band antenna may have only a single connection for the cable to your transceiver. If your transceiver uses only one port for both 2 meters and 70 centimeters, this isn't a problem. If the radio has separate input/output ports for each band, you'll need to use a device known as a *diplexer* to allow them to share the single cable to your antenna. See **Figure 4.1**. Diplexers typically sell for less than $100.

Speaking of cables, if you find that you must string more than 20 feet of coaxial cable between your transceiver and your antenna, be sure to use a low-loss cable such as LMR400. The losses at 430 MHz can be substantial, even with LMR400, so be sure to keep the cable as short as possible.

Option 2: Going Portable with a Gain Antenna

Let's add a gain antenna to the picture and move your operations outdoors. A gain antenna, such as a Yagi, will focus your transmitted energy and concentrate received signals as well. Lightweight dual-band gain antennas are commonly available. Two popular models on the market when this book was written include the Arrow 146/437-10WBP (**www.arrowantennas.com/arrowii/146-437.html**) and the Elk Antennas 2M/440L5 (**elkantennas.com/product/dual-band-2m440l5-log-periodic-antenna/**). You can also build your own gain antennas, as you'll see in Chapter 6.

The advantage of operating portable is that you can use a gain antenna without the need for expensive antenna rotators and many feet of coaxial cable and control wires. In a portable setup, *you* are the antenna rotator, turning and repositioning the antenna as necessary to keep the signal as strong as possible.

With a gain antenna, you won't need the higher RF output power used in the bare-minimum station. Many amateurs enjoy good results with just 5 W dual-band handheld transceivers. Of course, you can use more power, but be sure to check the power rating of the antenna first.

Some dual-band gain antennas provide separate coaxial cable connections for 2 meters and 70 centimeters. If this is the case with the antenna you've

The highly popular Arrow 146/437-10WBP is a dual-band Yagi that can be assembled and disassembled easily for portable satellite operating.

The Elk Antennas 2M/440L5 is a lightweight dual-band satellite antenna using a log periodic design.

chosen, you will need a diplexer, the same one used in the bare-minimum station option, to merge them into one cable connection for your transceiver.

The downside of using a portable antenna is that your hands and arms can become tired, even with the lightest antennas. Add the hassle of juggling your radio and perhaps a smartphone and you have a hectic operating experience. Hams who enjoy portable satellite operating usually end up adapting their antennas to camera tripods to make things easier!

Option 3: Upgrading the Home Station

Let's go back to Option 1. Once you're hooked on satellites, chances are you'll want to start upgrading that bare minimum home station. You don't need to upgrade all at once; doing it in steps is usually easier on your budget. See **Figure 4.2**.

The first item to upgrade is your antenna system. What is true for every other aspect of amateur radio is also true for satellite operating: *Your antenna system is the most critical component of your station*. If you are wondering how to invest your equipment funds, don't cut corners when it comes to purchases for your antenna system. An expensive, full-featured transceiver will be almost worthless if it is connected to poor antennas.

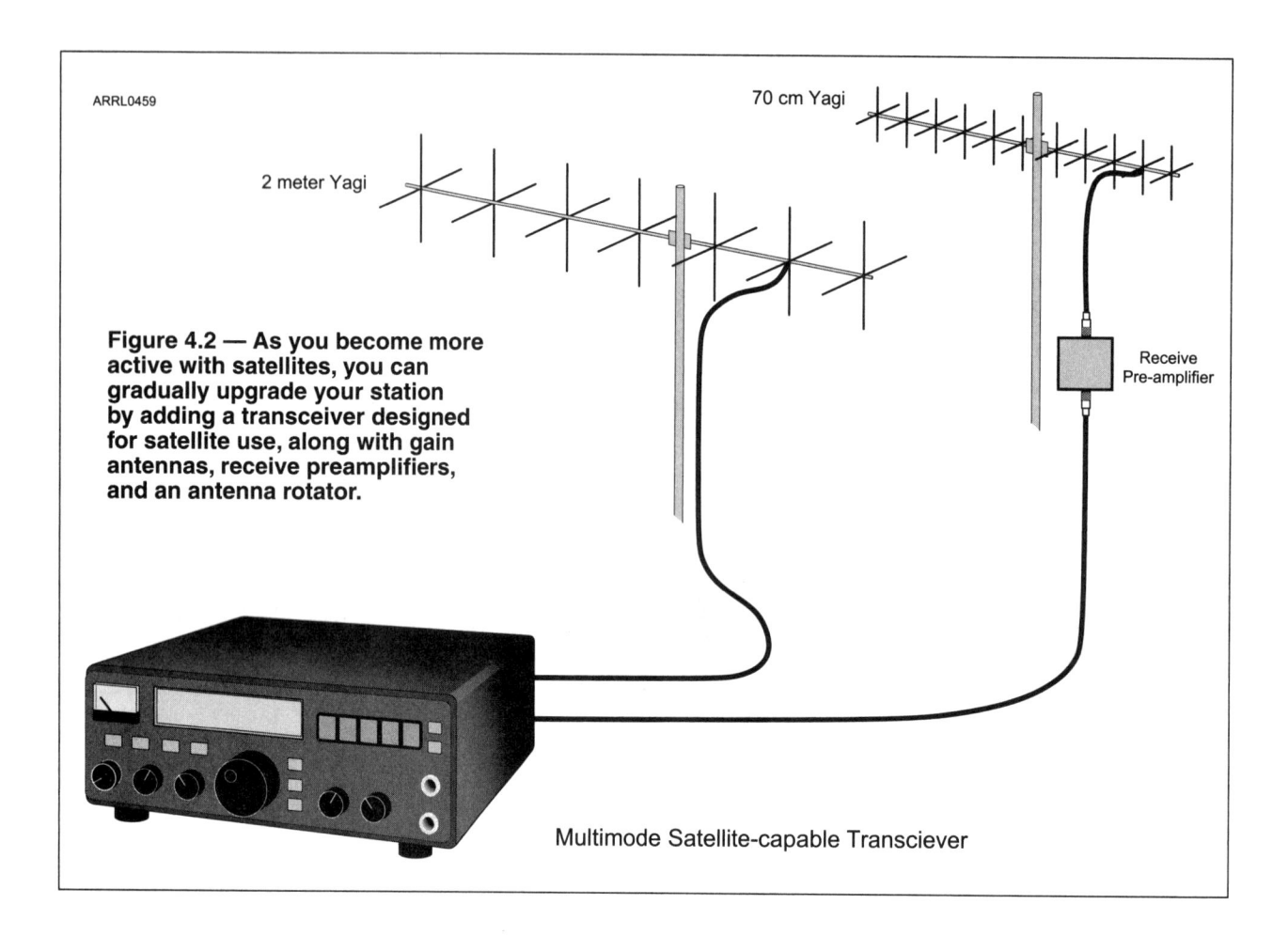

Figure 4.2 — As you become more active with satellites, you can gradually upgrade your station by adding a transceiver designed for satellite use, along with gain antennas, receive preamplifiers, and an antenna rotator.

And when we speak of the "antenna system," we're talking about more than just the antennas themselves. The "system" includes the feed lines that connect the antennas to your station. In the context of satellite operating, it may also encompass receive preamplifiers and antenna rotators.

The next step up from a dual-band omnidirectional base or mobile antenna is an omnidirectional antenna specifically designed for satellite use.

Omnidirectional Antennas

Despite their low gain, omnidirectional (omni) antennas specifically designed for satellite use can be viable options. Because they do not need to be aimed at their targets, they don't require mechanical antenna rotators, which can add significant cost and complication to a ground station. Omni antennas are also more compact than directional antennas for the same frequency. On the other hand, their low gain makes them practical only for low-Earth orbiting satellites that have sensitive receivers and relatively strong transmit signals.

For best results with an omni-based antenna system, you'll want a radiation pattern that minimizes the *pattern nulls* that can cause signals to fade. To this end, engineers have designed several omnidirectional antennas with these issues in mind. Note that you'll need two of the following antennas at your station — one for each band.

The Eggbeater Antenna

The eggbeater antenna is a popular design named after the old-fashioned kitchen utensil it resembles. The antenna is composed of two full-wave loops of rigid wire or metal tubing. Each of the two loops has an impedance of 100 Ω, and when coupled in parallel they offer an ideal 50 Ω impedance for coaxial feed lines. The loops are fed 90° out of phase with each other and this creates a circularly polarized pattern.

An eggbeater may also use one or more parasitic reflector elements beneath the loops to focus more of the radiation pattern upward. This effect makes it a "gain" antenna, but that gain is at the expense of low-elevation reception. Toward the horizon an eggbeater is horizontally polarized. As the pattern rises in elevation, it becomes more and more right-hand circularly polarized. Experience has shown that eggbeaters seem to perform best when reflector elements are installed just below the loops.

The eggbeater antenna is a popular design named after the old-fashioned kitchen utensil it resembles. The antenna is composed of two full-wave loops of rigid wire or metal tubing. To communicate with satellites on 2 meters and 70 centimeters, you'll need two eggbeaters — one for each band.

Eggbeaters can be homebrewed relatively easily, but there are also commercial models available, such as those made by M² (**www.m2inc.com**). The spherical shape of the eggbeater creates a compact antenna when space is an issue, which is another reason why it is an attractive design.

The Turnstile

The basic turnstile antenna consists of two horizontal half-wave dipoles mounted at right angles to each other (like the letter "X") in the same horizontal plane with a reflector screen beneath. When these two antennas are excited with equal currents 90° out of phase, their typical figure-eight patterns merge to produce a nearly circular pattern.

In order to get the radiation pattern in the upward direction for space communications, the turnstile antenna needs a reflector underneath. For a broad pattern it is best to maintain a distance of ⅜ wavelength at the operating frequency between the reflector and the turnstile. Homebrewed turnstile reflectors often use metal window-screen material that you can pick up at many hardware stores. (Make sure it is a metal screen as there is a nonmetal window-screen material as well.)

Like their cousins the eggbeaters, turnstiles are relatively easy to homebrew. In fact, homebrewing may be your only choice for turnstiles since they are rarely

The basic turnstile antenna consists of two horizontal half-wave dipoles, mounted at right angles to each other (like the letter "X") in the same horizontal plane with a reflector screen beneath. The antenna shown here is a commercial model for 2 meters sold by Wimo (www.wimo.com).

available off the shelf. One exception is a variation on the turnstile design — the model TA-1 for the 2-meter band sold by Wimo (**www.wimo.com**).

The Lindenblad Antenna

In a Lindenblad antenna, each dipole element is attached to a section of shorted open-wire-line, also made from tubing, which serves as a balun transformer. A coaxial cable runs through one side of each open-wire line to feed each dipole. The four coaxial feed cables meet at a center hub section where they are connected in parallel to provide a four-way, in-phase power-splitting function. This cable junction is connected to another section of coaxial cable that serves as an impedance-matching section to get a good match to 50 Ω.

The major cause of the difficulty in designing and constructing Lindenblad antennas is the need for the four-way, in-phase, power-splitting function. Since we generally want to use 50 Ω coaxial cable to feed the antenna, we must somehow provide an impedance match from the 50 Ω unbalanced coax to the four 75 Ω balanced dipole loads. Previous designs have used combinations of folded dipoles, open-wire lines, twin-lead feeds, balun transformers and special

The Lindenblad can be an excellent omnidirectional antenna for satellite use.

impedance-matching cables in order to try to get a good match to 50 Ω. These, in turn, increase the complexity and difficulty of the construction. In Chapter 6 you'll find a much easier Lindenblad design by the late Anthony Monteiro, AA2TX (who also provided some of the Lindenblad background details you've just read). While certainly more elaborate than an eggbeater or turnstile, the Lindenblad creates a uniform circularly polarized pattern that is highly effective for satellite applications.

Quadrifilar Helicoidal Antenna (QHA)

The quadrifilar helicoidal antenna ranks among the best of the omnidirectional satellite antennas. It is comprised of four equal-length conductors (*filars*) wound in the form of a corkscrew (*helix*) and fed in quadrature. The result is a nearly perfect circularly polarized pattern.

QHAs can be challenging to build since the filar lengths and spacing must be precise. Even so, homebrewing a QHA can save you a substantial amount of money. This antenna is available off the shelf (they are favorites for maritime satellite links), but they can be costly.

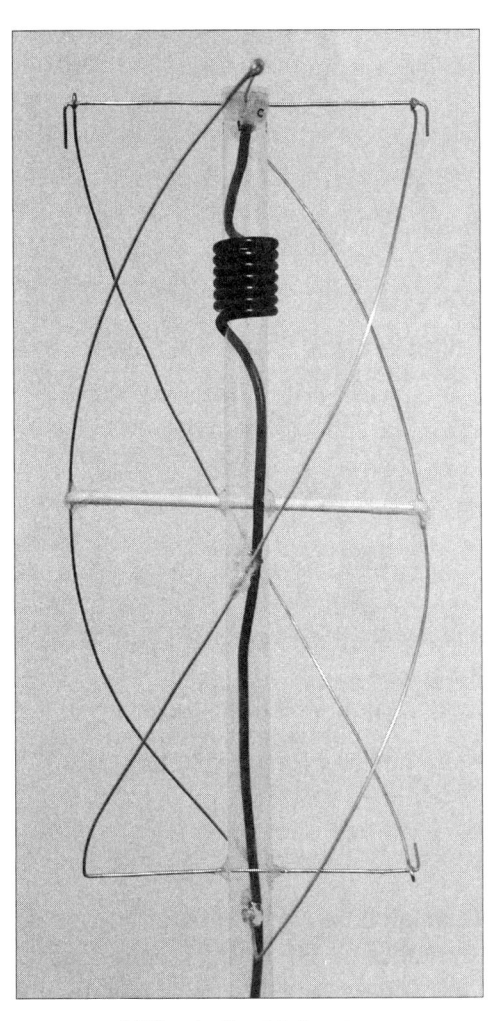

The quadrifilar helicoidal antenna is comprised of four equal-length conductors (*filars*) wound in the form of a corkscrew (*helix*).

Gain Antennas

The next rung on the home station upgrade ladder is an investment in gain antennas. But unlike the gain antenna we discussed in Option 2, this needs to be an antenna designed for permanent installation. You can choose a single dual-band antenna, or separate antennas for each band.

If you decided to invest in separate antennas for each band, the common approach is to mount them together on one horizontal boom. The boom should be made of a noninductive material such as thick-wall PVC, and the antenna should be positioned about 5 feet apart.

Before we get too far into the discussion of gain antennas, however, let's take a short detour and discuss polarization.

Antenna Polarization

The polarity of an antenna is determined by the position of the radiating element or wire with respect to the Earth, and it is an important factor when you are considering gain antennas.

A radiator that is parallel to the Earth radiates horizontally, while a vertical radiator radiates a vertical wave. These are so-called *linearly polarized* antennas and you'll find them for sale from several sources.

For terrestrial VHF+ line-of-sight communication, polarity matching is important. If one station is using a horizontally polarized antenna and the other

This is a dual-band (2 meters and 70 centimeters) Yagi antenna sold by Cushcraft. Notice that it is mounted in horizontal fashion, which will provide horizontal polarization. Notice also that it offers two separate feed lines. It isn't a satellite antenna per se, but it works quite well for communication with low-Earth orbiting birds.

is using a vertically polarized antenna, the mismatch can result in a large signal loss. We don't worry about polarization mismatches on HF frequencies because whenever signals are refracted through the ionosphere, as HF signals usually are, their polarities change anyway.

The problem with applying polarization concerns to spacecraft is that the orientation of a satellite's antennas relative to your ground station is constantly changing. This often results in fading when the polarities of its antennas conflict with yours.

Fortunately, there is a "cure" known as *circular polarization* (CP). With CP, the wavefront describes a rotational path about its central axis, either clockwise (right-hand; RHCP) or counterclockwise (left-hand; LHCP). The advantage of using circular polarization is that it can substantially reduce the effects of polarization conflict. Since the polarization of a CP antenna rotates through horizontal and vertical planes, the resulting pattern effectively "smoothes" the fading effects, generating consistent signals as a result.

That said, your gain antennas do *not* need to be circularly polarized to be effective! Linearly polarized antennas, either horizontal or vertical, are perfectly useful. Some ground station antenna designs use slanted or "crossed" elements to mix the horizontal and vertical polarization components as we discussed earlier. The goal of fine-tuning your antenna polarization is to give your station an edge, something that is important when you are dealing with weak signals from deep space. But, while circular polarization of your antennas may give your station a definite advantage, it isn't a requirement.

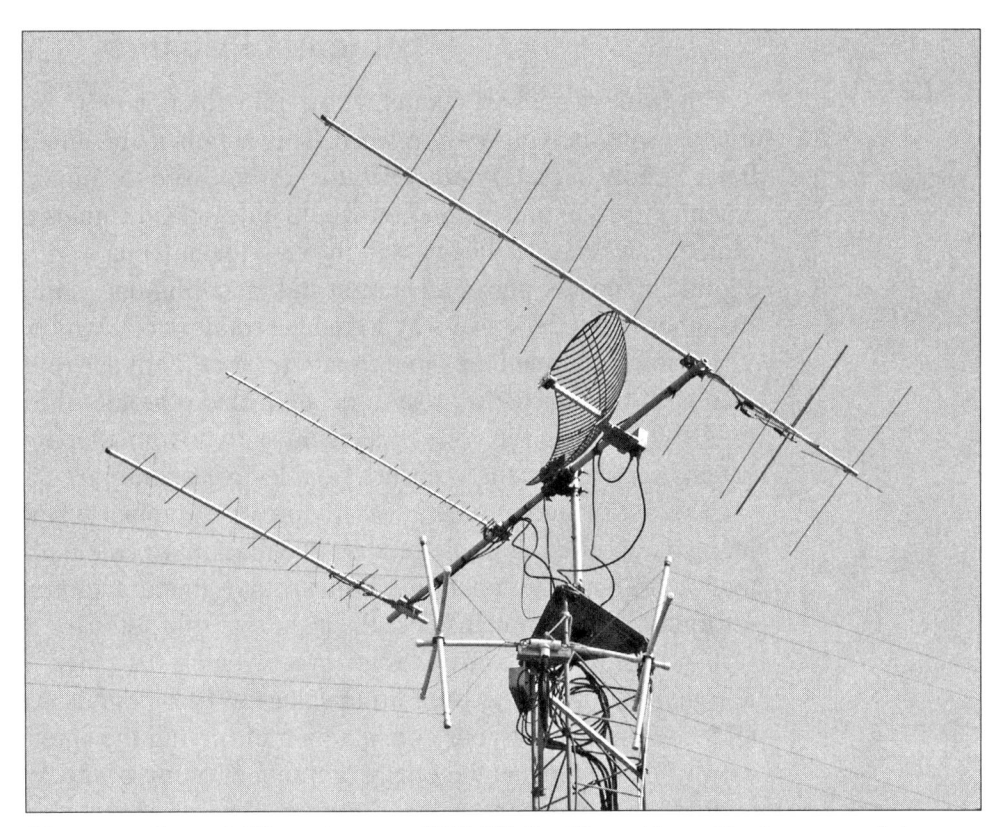

These are the satellite antennas at W1AW, the Hiram Percy Maxim memorial station at ARRL Headquarters in Newington, Connecticut. Notice the cross-polarized Yagi antennas for 2 meters and 70 centimeters attached to both ends of the horizontal support boom.

You'll find a few commercial dual-band cross-polarized antennas, but they tend to be somewhat expensive. If you can purchase two standard linearly polarized antennas, one for each band, chances are they'll provide adequate performance.

How Big Does a Gain Antenna Need to Be?

The most common gain antenna for satellite applications is the venerable Yagi. A Yagi can be short or long, big or small, depending on the frequency and the amount of gain you desire.

All antennas become smaller as the operating frequency increases. A Yagi antenna for the 70-centimenter band, for instance, can be easily half the size of the same antenna for 2 meters. The gain of a Yagi is dictated by the number of elements in the design and these elements also determine the total length.

Gain is a good thing, but as the saying goes, there is no such thing as a free lunch. When it comes to Yagi antennas, more gain means more elements, which means a longer antenna. But more than that, more gain means a narrower antenna pattern. A satellite can zip through a narrow pattern rather quickly, so you will be constantly shifting the positions of the antenna just to stay on target.

For low-Earth-orbiting satellites, my suggestion would be to stick with Yagi designs of six elements or less. They will give you enough gain to enjoy strong signals, while providing a broader radiation pattern.

Antenna Rotators

To achieve the best results with a gain antenna, you must find a way to point it at the satellite you wish to work. This entails using an electric motorized device known as an *antenna rotator*. If you were to simply leave a directional antenna fixed in one place, you would enjoy good signals only during the brief moments when satellites passed through the antenna's pattern. A rotator adds significant cost to a ground station and installing one isn't a trivial exercise. On the other hand, there is a way to reduce rotator cost, which we'll discuss later.

Rotators (some hams incorrectly refer to them as "rotors") are little more than high-torque electrical motors controlled remotely through a multiconductor cable. Making the correct decision as to how much capacity the rotator must have is very important to ensure trouble-free operation.

Rotator manufacturers generally provide antenna surface area ratings to help you choose a suitable model. The maximum antenna area is linked to the rotator's torque capability. Some rotator manufacturers provide additional information to help you select the right size of rotator for the antennas you plan to use. Hy-Gain provides an *effective moment* value. Yaesu calls theirs a *K-Factor*. Both ratings are torque values in foot-pounds. You can compute the effective moment of your antenna by multiplying the antenna turning radius by its weight. So long as the effective moment rating of the rotator is greater than or equal to the antenna value, the rotator can be expected to provide a useful service life.

There are several rotator grades available to amateurs. The lightest-duty rotator is the type typically used to turn TV antennas. These rotators will handle smaller satellite antennas such as crossed Yagis. The problem with TV rotators is that they lack braking or holding capability. High winds can turn the rotator motor via the gear train in a reverse fashion. Broken gears sometimes result.

The next grade up from the TV class of rotator usually includes a braking arrangement, whereby the antenna is held in place when power is not applied to the rotator. The brake prevents gear damage on windy days. If adequate precautions are taken, this type of rotator is capable of holding and turning a stack of satellite antennas, including a parabolic dish which, by its nature, presents considerable wind loading. Keep in mind that as rotators increase in power, they become more expensive.

The Azimuth/Elevation Rotator — Do You Really Need One?

Perhaps the ultimate in satellite operating convenience is the *azimuth/elevation* (az/el) *rotator*. This rotator can move your antennas horizontally (azimuth) and vertically (elevation) at the same time. There are well-designed models available from Yaesu and Alfa Radio. You can operate these rotators manually or connect them to your computer for automated tracking. The downside is that az/el rotators tend to be expensive, typically $700 or more (sometimes much more) at the time of this writing.

If your budget can stand the strain, az/el rotators are clearly worth the investment. On the other hand, if you're trying to shave pennies from your installation, consider using a standard rotator instead. While a traditional rotator can only move your antennas in the azimuth plane (horizontally), you can

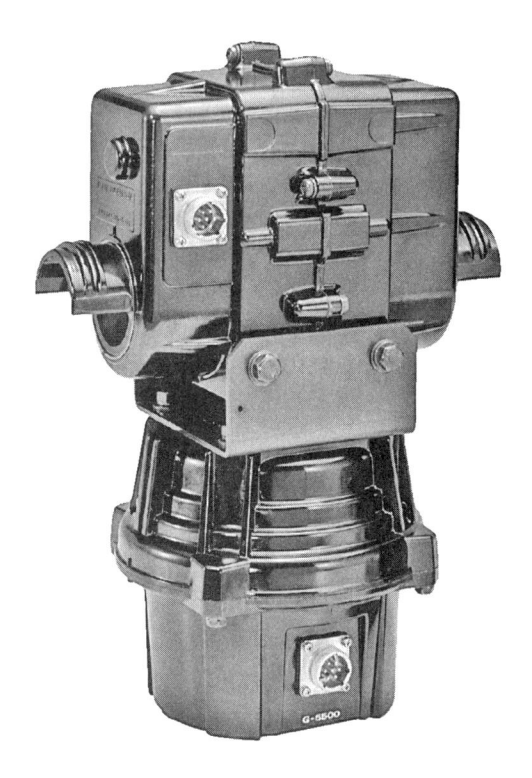

The Yaesu G-5500 azimuth/elevation rotator (left) and its control unit (right). At the time this book was written, it sold for about $700.

strike a compromise by installing the antennas at a permanent 30° tilt. Believe it or not, this configuration will allow you to work the vast majority of satellites with reasonable success. No, you won't be able to follow the satellite when it is overhead or near the horizon, but you'll enjoy the lion's share of every pass. Because you'll have pocketed several hundred dollars or more in the bargain, the loss in coverage may be worth the savings.

Regardless of which type you choose, proper installation of the antenna rotator can provide many years of dependable service. Sloppy installation can cause problems such as a burned-out motor, slippage, binding, and even breakage of the rotator's internal gear and shaft castings outer housing. Most rotators can accept mast sizes of different diameters, and suitable precautions must be taken to shim an undersized mast to ensure dead-center rotation. For instance, if you decide to install your rotator on a tower, it is desirable to mount the rotator inside and as far below the top of the tower as possible. The mast absorbs the torsion developed by the antenna during high winds, as well as during starting and stopping. Some amateurs have used a long mast that stretches from the top all the way to the base of the tower. This extreme example notwithstanding, a mast length of 10 feet or more between the rotator and the antenna will add greatly to the longevity of the entire system by allowing the mast to act as a torsion shock absorber. Another benefit of mounting the rotator 10 feet or more below the antenna is that the effect of any misalignment among the rotator, mast, and the top of the tower is less significant.

Don't forget to provide a loop of coax to allow your antenna to rotate properly. Also, make sure you position the rotator loop so that it doesn't snag on anything.

A problem often encountered in amateur installations is that of misalign-

> ## Antennas in the Attic?
>
> If you are living in a setting that doesn't allow outside antennas, you may want to look at your attic. Even apartments and condos often have attics accessible to the top-floor residents. Depending on the height of the attic, you may be able to install directional antennas and even antenna rotators. If the attic is small, consider omnidirectional antennas.
>
> The most serious problem with attic installations is the signal attenuation caused by roofing materials. Wood roofs with slate or asphalt shingles will pass VHF and UHF signals, but with considerable loss. Microwave signals will likely be attenuated to the point of being unusable. When the roof is wet, or ice and snow covered, both VHF and UHF signals will be further degraded. Metal roofs, of course, are deadly to all radio signals.
>
> For effective attic operation with low-orbiting satellites, receive preamplifiers are highly recommended. You may also find that you need to use more RF uplink power than you would with an outdoor antenna.

The Yaesu GS-232B allows your computer and satellite tracking software to automatically control the movements of compatible azimuth/elevation antenna rotators.

ment between the direction indicator in the rotator control box and the heading of the antenna. With a light duty TV antenna rotator, this happens frequently when the wind blows the antenna to a different heading. With no brake, the force of the wind can move the gear train and motor of the rotator, while the indicator remains fixed. Such rotator systems have mechanical stops to prevent continuous rotation during operation, and provision is usually included to realign the indicator against the mechanical stop from inside the shack. Of course, the antenna and mechanical stop position must be oriented correctly during installation. In most cases the proper direction is true north.

As mentioned earlier, you can connect your rotator to your station computer and allow your satellite-tracking software to aim your antennas automatically (assuming your software supports rotator control). There used to be several commercial rotator/computer interface devices available for sale, but availability has dwindled over the years and those that remain tend to be expensive. An interface such as the Yaesu GS-232B costs about $600. If you combine it with a Yaesu azimuth/elevation rotator, you will have invested close to $1,300. Less expensive alternatives are now found as kits or homebrew devices. A good example is the G6LVB tracker interface at **www.g6lvb.com/Articles/LVBTracker/**. If you're willing to build it yourself, you can probably put together an LVB unit for less than $100. Do a Google (**www.google.com**) search and you'll no doubt uncover other homebrew interfaces.

Receive Preamplifiers

Signals from satellites can be exquisitely weak, which means they need as much amplification as possible to be readable. Unfortunately, there are a few factors that may conspire to weaken your radio's ability to render a decent received signal...

You're using omnidirectional antennas. As we discussed earlier, omni an-

This receive preamplifier by Advanced Receiver Research gives signals a substantial boost before they travel down the coaxial cable to your radio.

tennas lack much of the signal-capturing gain of directional antennas.

The feed line between the antennas and the radio is long and/or contains "lossy" coax. Even with the best coax, the longer the feed line the more signal you'll lose, especially at higher frequencies.

The way to ensure that you have a useable received signal is to install a *receive preamplifier* at the antenna. This is a high-gain, low-noise amplifier with a frequency response tailored for one band only.

When shopping for a receive preamplifier, you want the most amount of gain for the least amount of noise. Every preamplifier adds some noise to the system, but you want the least additional noise possible. Preamplifier gain and noise are specified in decibels (dB). A well-designed UHF preamplifier, for example, may have gain on the order of 15 to 25 dB and a *noise figure* (NF) of 0.5 to 2 dB (less is better).

If your antennas are outdoors, look for preamplifiers that are "mast mountable." These preamplifiers are housed in weatherproof enclosures.

You will need to devise a means to supply dc power to the preamplifier. This can be as simple as routing a two-conductor power cable to the device. Alternatively, preamplifiers can be powered by dc sent up the feed line itself. Some transceivers can insert 12 V dc on the feed line for this purpose. If not, you can use a *dc power inserter* to inject power at the station and/or recover it nearer the antenna. Some preamplifier designs include feed line power capability, so all you need is an inserter at the "station end."

If your preamplifier is going to be installed in a feed line that will also be carrying RF power from the radio, you'll need a model that includes an internal relay to temporarily switch it out of the circuit to avoid damage to the preamplifier when you're transmitting. Look for preamplifiers that offer *RF-sensed switching*. This design incorporates a sensor that detects the presence of RF from the radio and instantly switches the preamplifier out of harm's way. Note that RF-switched preamplifiers are rated according to the power they can safely handle. If you're transmitting 150 W, you'll need an RF-switched preamplifier rated for 150 W or more.

Transceivers

There are many amateur transceivers that cover the VHF and UHF bands. If you only care about enjoying the FM repeater satellites, a dual-band FM transceiver is all you need. But if you intend to expand your horizons to include operating CW and SSB, you will need a more elaborate radio that includes features specifically intended for satellite operating.

FM signals tend to be wide and, by design, FM receivers are forgiving of

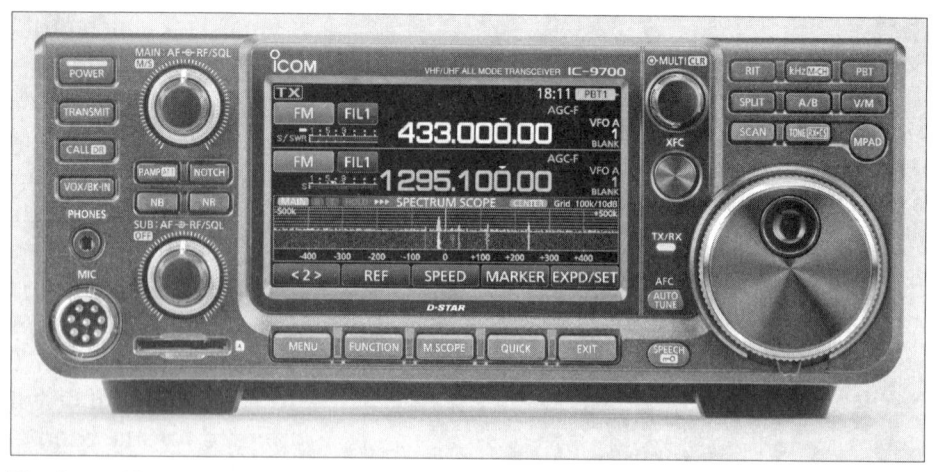

The Icom IC-9700 is an all-mode VHF/UHF transceiver with dedicated satellite functions, including full duplex. It was released in 2019.

frequency changes. That fortunate characteristic makes it easy to compensate for Doppler frequency shifting as an FM repeater satellite zips overhead. As you'll learn in the next chapter, you can program transceiver memory channels with a few uplink/downlink frequency steps and switch from one to another during the pass.

On the other hand, SSB and CW signals are much narrower, and when you're working through a linear transponder satellite your signal is sharing the passband with several others. Not only do you need to adjust your transceiver frequency almost continuously to keep the SSB voice or CW sounding "normal," you also must stay on frequency to avoid drifting into someone else's conversation. The most effective way to do this is to listen to your own signal coming through the satellite in "real time" while you are transmitting on the uplink. This type of operation is known as *full duplex*.

When it comes to considering an SSB/CW transceiver for satellite operating, the issue of full duplex capability is potentially confusing. You will find many multimode (SSB, CW, FM) transceivers that boast a feature labeled "cross-band split" or even "cross-band duplex." Be careful, though. What you

The Kenwood TS-2000 offers full HF coverage, all modes, plus VHF and UHF with full-duplex satellite capability. It is no longer in production but can still be found on the used market.

require is a radio that can transmit and receive on different bands *simultaneously*. Few amateur transceivers can manage such a trick!

As this book went to press, there was only one VHF/UHF all-mode, full-duplex, amateur transceiver in current production: the Icom IC-9700. However, if you shop the used equipment market, you'll find excellent satellite radios such as the Kenwood TS-2000, the Yaesu FT-736, the Icom IC-820 and IC-910, the Yaesu FT-726 and FT-847, along with the Kenwood TS-790. All these transceivers have full-duplex capability.

VHF/UHF RF Power Amplifiers

If your chosen transceiver offers at least 50 W output on the uplink band, you won't need an RF power amplifier to bring your signal to a level that can be "heard" by a low-Earth orbiting satellite, especially if you are using directional antennas.

If you are using omnidirectional antennas, 100 or 150 W output may help considerably. If your transceiver lacks the necessary punch for the application, the solution is an external RF power amplifier.

How much power should you buy? In most cases, a 100 or 150 W amplifier is a good choice. As you shop for amplifiers, take care to note the input and output specifications. How much RF at the input is necessary to produce, say, 150 W at the output? Can your radio supply that much power?

Another consideration is your dc power supply. While a 25-A 13.8 Vdc supply is perfectly adequate to run a 100 W transceiver, if you also decide to add a 100 or 150 W amplifier to your satellite station, the current demands will increase considerably. A separate power supply may be required to provide an *additional* 20 A (or more) to safely power the amplifier when both the transceiver *and* the amplifier are transmitting at the same time.

Chapter 4

Let's Get on the Air!

We've spent the last three chapters talking about satellite history, satellite tracking, and satellite equipment. Now it's time to switch on your radio and start making contacts.

You might expect this chapter to begin with a list of all the available satellites and their frequencies, however, old satellites die and new satellites reach orbit at a rate far faster than any book could possibly cover. A satellite that is active this week may not be active next week. And by the time this book rolls off the press, it is likely that a new satellite will join our amateur radio "space fleet." A year or two later, there will be even more.

In this rapidly changing environment, one internet information source is constant: AMSAT-NA at **www.amsat.org**.

The AMSAT Live Oscar Satellite Status Page (**www.amsat.org/status/**) provides a list of every currently active satellite. Best of all, the status table will tell you if the satellite is presently on the air, if its transponder (repeater) is available, or whether the satellite is just a telemetry-only bird. You can determine when it was last heard, as well as who reported hearing it.

Another excellent AMSAT resource to determine the status (and frequencies) of many satellites is **www.amsat.org/two-way-satellites/**.

While you are at the AMSAT site, take a moment to sign up for membership. Most of the satellites you'll enjoy are the creations of unpaid AMSAT volunteers throughout the world. By joining AMSAT, you will lend your support to keep more satellites coming our way in the future.

The AMSAT satellite status page at www.amsat.org/status/, as it appeared in late 2019.

FM Repeater Satellites

Regardless of how you've equipped your station, an FM repeater satellite is almost always within radio reach. Pick your bird from the AMSAT status page and, as we discussed in Chapter 2, use your software, smartphone app, or the AMSAT Pass Prediction web page to determine when the spacecraft will be visiting your area.

If you are using the bare-minimum station with an ordinary omnidirectional antenna, look for a pass that puts the satellite at least 60° above your local horizon at highest point of the pass. If you have directional gain antennas, you can attempt passes with much lower maximum elevations. In fact, if you live on flat terrain with few hills or other obstacles, you may be able to contact satellites when they are barely above the horizon.

While you are picking your pass, don't forget to look up your grid square, as we also discussed in Chapter 2. You'll need it when the action starts!

FM repeater satellites tend to be very crowded, the VHF/UHF equivalent of DX pileups on the HF bands. If you are using the bare-minimum station, or gain antennas and low output power (such as a portable antenna with a 5 W handheld transceiver), try to pick a daytime pass during a weekday, or perhaps a late-night pass. Weekend operating can be brutal, but if you have significant RF output power and gain antennas, you can still cut through the tsunami of signals.

Memories Are Your Friends

Every modern FM transceiver has frequency memories, and these can be lifesavers when you need to compensate for the effects of Doppler shift. Pick a couple of FM repeater satellites that you think will be good candidates and

Adam Whitney, KØFFY, made contacts through the AO-92 FM repeater satellite while standing on the rim of the Hverfjall volcano crater in northern Iceland.

write down their uplink and downlink frequencies.

When this book was being written, one of the most popular FM repeater satellites was AMSAT-OSCAR 91, or simply AO-91. With the hope that AO-91 is still active when you read this, let's use it as our programming example.

AO-91 listens on 70 centimeters (its *uplink*) and transmits on 2 meters (its *downlink*). With that in mind, program five memory channels and label them as follows (assuming your radio offers memory labeling):

Label	Transmit Frequency	Receive Frequency
AO91 AOS	435.250 MHz	145.970 MHz
AO91 APP	435.250 MHz	145.965 MHz
AO91 MID	435.250 MHz	145.960 MHz
AO91 DEP	435.250 MHz	145.955 MHz
AO91 LOS	435.250 MHz	145.950 MHz

Also, program each memory channel to transmit a 67 Hz CTCSS tone (just like an FM repeater here on Earth).

By programming these memories, you can easily change the receive frequency of your transceiver so you can compensate for the Doppler shift. Thanks to the wide FM bandwidth of the satellite receiver, you do not need to change your transmit frequency.

Let's Get on the Air!   4-3

The Central Mississippi Contest Group made satellite contacts for bonus points during the 2019 ARRL Field Day.

The pass begins at AOS (Acquisition of Signal) when the satellite rises high enough above the horizon to allow you to start hearing signals.

APP is the approach, when the satellite is above the horizon and heading your way.

MID is the middle of the pass, when the satellite is at its highest point.

DEP is departure, as the satellite begins to sink back down toward the horizon.

LOS is Loss of Signal, when the bird drops so low that you'll soon lose its signal.

Notice how the receive signal frequency decreases by 5 kHz at each step. You can use this same amount of frequency decrease with other satellite memories, regardless if the uplink is at 2 meters or 70 centimeters.

Program the memories for each satellite as follows:

- Program the AOS memory to receive **10 kHz above** the published satellite transmit (downlink) frequency and to transmit on the satellite's published receive (uplink) frequency.
- Program the APP memory to receive **5 kHz above** the published satellite transmit (downlink) frequency and to transmit on the satellite's published receive (uplink) frequency.
- Program the MID memory to receive directly at the published satellite transmit (downlink) frequency and to transmit on the satellite's published receive (uplink) frequency.
- Program the DEP memory to receive **5 kHz below** the published satellite transmit (downlink) frequency and to transmit on the satellite's published receive (uplink) frequency.
- Program the LOS memory to receive **10 kHz below** the published satellite transmit (downlink) frequency and to transmit on the satellite's published receive (uplink) frequency.

As with AO-91, if the satellite repeater uses CTCSS, you will need to program each memory to transmit the proper tone.

Let the Games Begin

As the time for the beginning of the pass approaches, keep an eye on your software. If your software offers a footprint display, you'll see the edge of the footprint coming closer and closer. You'll have only 10 or 15 minutes, at most, to make contact, so be prepared!

Fire up your transceiver, switch to your AOS memory channel, and turn off your squelch. If you have an audio recorder, or a smartphone with an audio recording app, this would be a good time to turn it on. By recording the audio, you can avoid the hassle of scribbling call signs.

If you are using a directional gain antenna, aim it at the point where you expect the satellite to rise above the horizon. Listen carefully. As the satellite rises, one of the first things you may notice is a gradual reduction in receiver noise. That's the "quieting" effect of the satellite's FM signal in your radio.

Be patient and keep listening. It won't be long before you begin to hear hints of voices among the noise. *Do not start transmitting.* If you can't hear the other stations clearly, chances are they won't hear you clearly either. If you start calling, you will cause interference for everyone else.

When the signals become clear, switch to your APP memory. Now it's time to transmit. Don't expect an orderly exchange of contact information; this will be unadulterated chaos. The best you can do is listen for the smallest break in the action and make your play, like this:

"WB8IMY, Fox Nancy 31" (My call sign and my grid square, spoken phonetically.)

If I don't hear a response, I'll wait for the next opportunity and call again. Maybe next time, I'll be lucky.

"WB8IMY, Fox Nancy 31."
"WB8IMY, this is N9ATQ, Echo Mike 87."
"N9ATQ. Thanks for the contact!"

If you are using a gain antenna, you will have to move the antenna with the satellite as it races across the sky. At the same time, occasionally glance at a clock or your software display. When the satellite reaches the midpoint of the pass, switch the transceiver to your MID memory channel. You will know it is time to adjust your antenna aiming when you hear the signal beginning to fade.

When you make contact, resist the temptation to start a conversation. That is considered a rude practice on FM repeater satellites. Instead, wait a minute and try again, or try answering someone else who is throwing out their call sign and grid.

When the bird starts sinking toward the horizon, switch to the DEP memory channel. This is still a good time to make contacts. As the satellite's footprint extends away from your location, it includes stations that you would not have been able to contact before. If you have a directional antenna, switch to the LOS memory and try following the bird all the way down to the horizon.

Sometimes the action is so frantic, it is over before you know it. The satellite sinks back into the static and is gone. It is time to write down the call signs, grid squares, and the approximate times that the contacts occurred.

If you have logging software, enter the contacts and upload them to ARRL's Logbook of The World. See the sidebar, "Satellite Contacts and LoTW."

Satellite Contacts and LoTW

John Barber, N5JB

You can enter your satellite contacts into your logging software and then upload them to ARRL's Logbook of The World (LoTW) to qualify for awards such as the VHF/UHF Century Club. However, you will need to take several steps to prepare the output file for uploading.

Let's assume that you are a satellite user and that you have taken all the steps to prepare to use LoTW for the first time as detailed in the "Get Started" section of the ARRL LoTW web page, and are now ready to start the process of uploading your satellite log.

Follow your logging software's instructions to output an ADIF file that contains the satellite contacts you want to upload to LoTW. I recommend that you output it to your desktop, then open the ADIF file in Windows Notepad or another word processing application that can generate plain text files.

A properly arranged and formatted ADIF file is the secret to successfully uploading your satellite contacts to LoTW. Your logging program will output lines of data including RST, QSL info, and other lines of data, much of which will be ignored by the TQSL program. A DX or domestic contact record output from your logging program to an ADIF file may look something like this:

```
<CALL:5>KH6BB <QSO_DATE:8:D>20080608 <TIME_ON:6>180108
<TIME_OFF:6>180151
<FREQ:8>14.26300 <BAND:3>20M <MODE:3>SSB <TX_PWR:2>KW
<RST_SENT:2>56 <RST_RCVD:2>59 <QSL_SENT:1>N <QSL_RCVD:1>N
<NAME:25>BATTLESHIP MISSOURI AMATE <QTH:4>AIEA <STATE:2>HI
<COMMENT:18>USS MISSOURI - JIM
<GRIDSQUARE:4>BL11

<CNTY:11>HI,HONOLULU
<CQZ:2>31
<PFX:3>KH6
<DXCC:3>291
<OPERATOR:4>N5JB
<EOR>
```

It looks intimidating, doesn't it? Once boiled down to the five essential lines of data in the ADIF record of your non-satellite contacts, it will look like this:

```
<CALL:5>KH6BB
<QSO_DATE:8:D>20080608
<TIME_ON:6>180108
<BAND:3>20M
<MODE:3>SSB
<EOR>
```

If you upload the entire record (and you probably will), LoTW will ignore all but these five essential lines of data. In other words, you do not need to edit out the nonessential data. LoTW will do that for you by picking out only what it needs. Your "boiled down" satellite contact records will need to include the two additional data lines in order to successfully build a satellite contact record in LoTW:

```
<CALL:6>WA6DIR
<QSO_DATE:8>20080710
<TIME_ON:6>151200
<BAND:4>70CM
<MODE:3>SSB
<PROP_MODE:3>SAT
<SAT_NAME:4>AO-7
<EOR>
```

Notice the difference between a DX record and a satellite record. You can see that the satellite record includes two additional lines of data:

Propagation Mode: <PROP_MODE:3>SAT

Satellite Name: <SAT_NAME:4>AO-7

The ADIF contact record output from your logging program may contain some additional lines of data such as:

<FREQ:7>432.150
<BAND_RX:2>2M
<FREQ_RX:5>145.9

They may look like they ought to be essential, but they are not needed to successfully upload a satellite contact record to LoTW. You can leave them in or edit them out.

If your logging program did not output lines of data for the Propagation Mode and the Satellite Name, you will need to edit the ADIF satellite contact records to add them. The best way is to copy and paste using Notepad, or you can use the TQSL editing tool. To edit the ADIF file using the TQSL editor, open TQSL and click on "edit existing ADIF file."

It is much more time consuming to use the TQSL editor than to copy and paste into Notepad. In Notepad you can type the changes to one of the ADIF records, then copy and paste to all other records to which the same data lines apply. You can open Notepad (or the TQSL editor) and your logbook on your computer screen and switch back and forth between them. After you make the appropriate changes, resave ADIF file.

After you get your ADIF file into proper form, you are ready to sign and upload your log to LoTW. First use TQSL to sign and convert your edited ADIF file into a .tq8 file. LoTW will not upload the ADIF file.

Sign on to LoTW and click on UPLOAD. Open the .tq8 file you just created and upload this file. LoTW will do the rest. You can check on its progress and the results by going to YOUR ACCOUNT and looking at the ACTIVITY file. It will tell you if your upload was successful and if there were any contacts rejected for some reason. Give it a little while (usually a few minutes but occasionally a few hours) and then check to see if your contacts are there and if there have been any "hits."

Some logging programs allow you to redefine fields. For instance, the *WinEQF* program allows you to redefine the TRACK and the INFO fields. You could redefine them as the Propagation Mode and the Satellite Name fields so that the program will output the proper ADIF data. Note, however, that you cannot use these fields for any other purposes or add any extra data in the field or it will confuse LoTW and the record will be rejected. You will still have to add any missing fields to the ADIF file. Note that a number follows the colon inside the < > brackets. This number refers to the number of characters needed for the data. For example, the four in SAT_NAME:4>AO-7 means that the AO-7 satellite designator requires four characters in the data field. You may have to adjust this number according to the size of the data field.

LoTW wants the Satellite Name entered exactly as on the list of accepted satellites. For instance, if you enter the satellite name as AO7 instead of AO-7 the data will be rejected during the upload. You can find the most current list online at **https://lotw.arrl.org/lotw-help/frequently-asked-questions/#sats**.

The International Space Station

The International Space Station (ISS) is equipped with a powerful FM station, all thanks to *ARISS*: Amateur Radio on the International Space Station (**www.ariss.org**). It is one of the most powerful signals you are likely to hear. Even with a bare-minimum station, you'll be surprised at how "loud" the ISS can be.

The ISS orbits at a much lower altitude than most satellites, which means that its footprint is considerably smaller than the FM repeater satellites we've been describing. As a result, the activity windows during ISS passes may open and close within just 10 minutes.

Much of the ARISS activity involves prearranged contacts between school students and the astronauts and cosmonauts aboard the station, which we'll discuss in Chapter 5. However, the ISS is also active in other ways.

At certain times, the station will transmit Slow Scan TV (SSTV) signals on 145.800 MHz. You can receive these images using your FM transceiver by routing the received audio to your computer sound device input and using software such as *MMSSTV* (**https://hamsoft.ca/pages/mmsstv.php**). If you have a smartphone or tablet, you can decode the images by using an app such as SSTV Pad (for iOS) or DroidSSTV (for Android) and holding your device close to the radio speaker.

The ISS also offers a packet radio repeater and it is often active on 145.825 MHz. Packet radio is a form of digital communication that you can use to exchange short bits of text, as well as your location, using the Automatic Position Reporting System (APRS).

A thorough description of packet radio and APRS is beyond the scope of this book; entire books have been written on these topics. With that said, it is not as complicated as it may seem.

If you are willing to invest in a more expensive dual-band FM transceiver, you will find models that contain all the necessary packet radio hardware and APRS software. If you happen to own one of these radios, all you have to do is tune to 145.825 MHz during an ISS pass and you will see packet activity. To join the fun, especially with APRS, you only need to change a few settings in your radio. Set the UNPROTO command to CQ VIA ARISS, for example. You can add additional routing later as you become more familiar with packet operations.

If your transceiver doesn't include packet radio capability (most don't), you will need to invest in a bit of external hardware and software. See this page on the web for more information: **https://issfanclub.eu/2019/04/29/aprs-via-iss-tips-for-successful-operation/**.

There is a voice repeater aboard the station, but it is usually inactive. On rare occasions the crew will find moments in their busy schedules to grab the microphone and make contacts. Just like DX pileup on the HF bands, the crew uses a split-frequency approach:

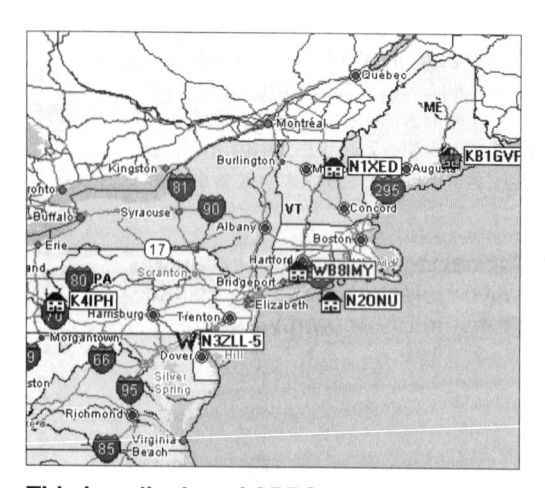

This is a display of APRS stations that had position packets repeated by the ISS during a pass over the eastern United States.

they listen on one frequency and transmit on another. When the station is over the Americas, you will need to transmit on 144.490 MHz and listen on 145.800 MHz.

Once again, this activity is very uncommon and can happen without notice. Occasionally, during special events such as ARRL Field Day, the crew will announce their intentions in advance. Keep an eye on the ARRL website (**www.arrl.org**) for announcements.

See Chapter 5 of this book for even more detailed information about ARISS and the International Space Station.

Linear Transponder Satellites

Some amateurs refer to FM repeater satellites as "EasySats," and for good reason. From an operational point of view, they are essentially FM repeaters in outer space. Once you work through the tracking and equipment issues, making a contact through an FM repeater satellite is fundamentally the same as chatting with a buddy through a terrestrial repeater while driving home from the office — just a lot shorter.

Linear transponder satellites also repeat signals, but they are very different. Making contacts through these birds requires different equipment and different skills.

Some may say that a book devoted to beginner-level satellite operating should omit linear transponder satellites entirely. I disagree. Yes, the path to success with these satellites is steeper; it is also more expensive. But some day you will grow weary of trading grid square designators on the FM birds and you'll want to have real conversations. This is what a linear transponder satellite offers. It also offers the satisfaction that comes with tackling a greater challenge.

What is a Linear Transponder?

Consider the 40-meter HF band from 7.000 to 7.300 MHz. At the low end of the band, you have CW signals. As you move beyond about 7.070 MHz, you encounter digital activity. Go higher still and you will hear SSB and AM voice conversations, along with the occasional shortwave broadcast signal.

Now imagine a device that could take every signal heard between 7.000 and 7.300 MHz and retransmit all of it within the same 300 kHz range *at completely different frequencies*. For example, imagine taking everything you can hear between 7.000 and 7.300 MHz and retransmitting it between 29.000 and 29.300 MHz.

Such a device exists: the linear transponder.

Let's look at a real-world example with the Chinese XW-2A satellite. This satellite has a linear transponder that listens between 435.030 and 435.050 MHz. This 20 kHz range is called its *uplink passband*. Everything XW-2A hears in that frequency range is retransmitted between 145.665 and 145.685 MHz. That is called its *downlink passband*.

Life with linear transponders would be easier if they performed direct, one-for-one, frequency conversions. Unfortunately, for technical reasons that we won't get into here, most do not. Instead, they usually *invert* the signal frequencies from the uplink to the downlink passbands. A signal transmitted to the satellite at, say, 435.035 MHz (5 kHz *above the low end* of the uplink passband) will appear at about 145.680 MHz (5 kHz *below the top end* of the downlink passband). In addition, a signal that is lower sideband on the uplink will be retransmitted as upper sideband on the downlink.

Now you know why linear transponder operating presents such a challenge! You must think in a completely different way when it comes to choosing your transmitting and receiving frequencies.

Why Only SSB and CW?

FM signals are forbidden on linear transponder satellites. Why is that?

Part of the answer is bandwidth. A typical SSB signal has a bandwidth of about 2800 Hz. When you consider the transponder bandwidth of a satellite such as XW-2A, nearly 10 SSB conversations could squeeze into the available space. A CW signal is much narrower, so a linear transponder can accommodate many more of those. Considering the bandwidth of your average FM signal, however, only a couple could "fit" within the XW-2A passband.

Perhaps the most important part of the answer is RF power. When a linear transponder retransmits its downlink passband, every signal gets a share of the total RF output. So, if a transponder has only 2 W of total output available, that power must be shared among all signals. The wider the bandwidth of a given signal, the more power it requires. A relatively narrow CW or SSB signal requires much less downlink passband power than a wide FM signal. Yes, a linear transponder can retransmit an FM signal, but to do so it would starve all the other signals of the power they need to be heard.

An FM transmission also takes place at a 100% *duty cycle*, which means the transmitter operates at 100% of the selected power level throughout the entire transmission. In contrast, an SSB transmission takes place with an average duty cycle of only 30%. Spacecraft power systems must operate within tight limits and are easily overtaxed. Generating full power continuously for several minutes to support an FM transmission would be too much to ask of an amateur radio satellite.

Equipment for Linear Transponder Operating

In Chapter 3, we discussed the so-called "upgraded" satellite station. This is the type of station you will need to be effective with linear transponder satellites.

You can use omnidirectional antennas to communicate through linear transponder satellites, but the results may be disappointing. You can mitigate some of this by using an RF power amplifier on the uplink and receive preamplifiers on the downlink.

Even so, directional gain antennas will work much better, even if they are at a fixed 30-degree elevation with a conventional TV antenna rotator. You'll be dealing with weaker signals at times, so a receive preamplifier is always a good idea.

A close-up image of a Kenwood TS-2000 transceiver display while operating in the satellite mode. It is transmitting on the 70-centimeter band while listening in full duplex on the 2-meter band. Notice that the Trace function is active.

The single item that will do the most to preserve your sanity is a transceiver that offers features designed for satellite operating. The Kenwood TS-2000 and the Icom IC-9700 are good examples. Both offer *full duplex* — the ability to transmit and receive simultaneously. This is a required feature for linear transponder operating, as you'll see later.

Even better, these radios include the ability to "track" (or "trace") frequency tuning between two bands, even if you are grappling with an inverting transponder. With inverted tracking enabled, for example, as you tune your transmit frequency *upward*, the radio will automatically tune your downlink frequency *downward*, or vice versa.

These advanced radios can also communicate with your station computer, allowing the computer to set the frequencies and change them automatically to compensate for Doppler shifting. It is almost like having a second human operator at your station!

Linear Transponder Operating Tips

Use the AMSAT status page mentioned earlier in this chapter to find an active satellite and determine its uplink and downlink passband frequencies.

As the satellite rises above your horizon, start tuning through the downlink passband and listening for signals. You might also try listening for the satellite's beacon (if it has one).

When you begin hearing signals, try some test transmissions, using either CW or SSB. With an inverting transponder, transmit using lower sideband and listen using upper sideband.

Plug in a pair of headphones and, with the transceiver in the full duplex mode, estimate where your signal should appear in the downlink passband. Begin transmitting and listen for your own signal being retransmitted by the satellite. It is always exciting to hear your own voice or CW signal coming back from outer space!

As you listen, compensate for Doppler by adjusting the tuning for the *higher* frequency band to keep your voice sounding normal. For example, if the uplink passband is on 70 centimeters while the downlink passband is on 2 meters, you'd adjust the uplink frequency, which is on the higher band, while keeping your downlink frequency untouched. On the other hand, if the uplink passband is on 2 meters and the downlink passband is on 70 centimeters, you would leave your uplink frequency alone and instead adjust your downlink frequency.

Let's Get on the Air!   4-11

Start calling CQ and keep compensating for Doppler. Doppler shifting becomes more pronounced as you increase frequency, so you will quickly discover that a 70-centimeter downlink signal requires almost constant tweaking. This is where a computer comes in handy.

When someone responds, you'll probably hear them "swooping in" to your frequency. Their voice or CW signal may have a strange pitch that will quickly change to match your pitch.

Once you've established contact, keep compensating and start talking. Unlike the FM repeater birds, the only constraint on your time is how long the satellite remains in range. Some of the higher-orbiting satellites will remain in range for up to 20 minutes.

Chapter 5

Satellites and Education

You don't have to be a teacher to use amateur satellites as educational tools. Spacecraft have powerful roles to play in formal education, but they can "teach" in other situations as well.

Using Satellites to Promote Amateur Radio

How many times have you mentioned amateur radio to someone, only to hear, "Do people still do that?" or, "Don't you mean CB radio?"

Hams have a public relations problem. People either believe we no longer exist, or they confuse us with other radio services. Worse yet, there are those who know that ham radio still exists, but they believe it is a hobby for old men who play with ancient technology in their basements.

Satellites offer an excellent means to shatter these musty stereotypes. Satellite demonstrations can be impressive. There you are, pointing an antenna at an invisible object in the sky, and voices suddenly begin streaming from the radio in your hands. You turn to your audience and announce, "What you are hearing is a signal from a spacecraft streaking 800 miles above our heads at more than 17,000 miles per hour!"

You don't even need to make a contact, although that would sweeten the deal. Your pitch could be something as simple as, "You don't need the internet to talk to one of these spacecraft. This is called amateur radio!"

Of course, the trick is finding an active satellite that's due to pass overhead at an appropriate time. If you are conducting a demonstration at an all-day event such as a town fair, you'll almost certainly enjoy at least one high-elevation pass. If you are considering a shorter event, such as a one-hour seminar at your

Dan, AC1EN, and Mackenzie, KE1NZY, explain amateur radio satellites to a non-ham audience.

local library, you'll need to consult your pass-prediction software and schedule the seminar for a date and time that coincides with a pass.

I strongly recommend the use of a wide-screen monitor connected to a laptop computer. Show your audience the satellite they are about to hear and the path it is taking to your location. A software display in tandem with a live demo can be riveting. The audience sees the spacecraft approaching and they begin to hear its signals in your radio. Even in the 21st century, it almost seems like science fiction.

If you can't do a live demo, don't worry. Make audio recordings of passes and use them instead. Use the big screen monitor and software. Show your audience all the amateur radio satellites presently in orbit. You can toggle your software to jump from one bird to another. Be sure to remind them that what they are seeing are the movements of real spacecraft. Tell them about Amateur Radio on the International Space Station (ARISS) as well.

At your demonstrations, be prepared to answer questions about how individuals can obtain amateur radio licenses. Have printed handouts available with

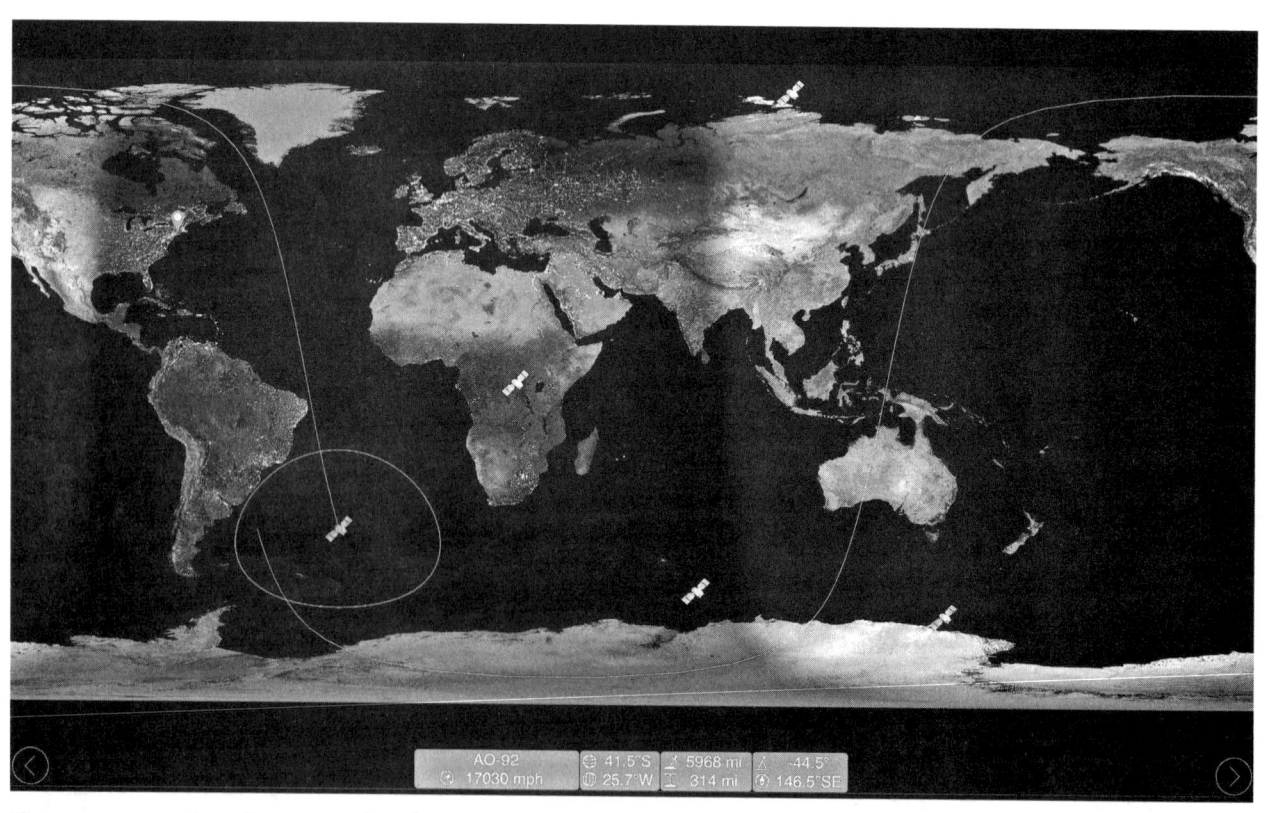

If you can acquire a large monitor for your demonstration, use it to display the orbital track of the satellite the audience is about to hear.

5-2   Chapter 5

more information. It is one thing to entertain, but you need to "close the sale" as well. If you have a local club that can offer license classes to coincide with your demonstration, so much the better.

Satellites in the Classroom

If you are a teacher, amateur satellites can play a powerful role in science instruction. Amateur satellite software can provide vivid displays of satellites in various orbits, and the orbital paths can be viewed in the present and the future, demonstrating how orbits can change over time.

Of course, teachers do not need ham licenses to monitor satellite passes. Imagine the educational impact of selecting a pass prediction for a given date and time, and then going outdoors with a portable gain antenna and a handheld transceiver to hear the spacecraft as it zips overhead.

Aside from satellite-tracking software and a large display monitor, a permanently installed outdoor gain antenna, with an azimuth/elevation rotator, would be ideal. If that investment is too much for the school's budget, the outdoor portable option is entirely acceptable. The goal is to engage students in understanding orbital mechanics at a fundamental level, as well as the physics behind radio communications.

The FUNcubes

Since 2013, several satellites have been launched specifically for educational outreach (although they also have transponders for normal amateur radio enjoyment). They were sponsored by AMSAT-UK and AMSAT-NL and are known worldwide as the *FUNcubes*.

The first of these was FUNcube-1, also known as AO-73, which was still active when this book went to press. During the sunlight portions of its orbit, FUNcube-1 transmits telemetry that details conditions aboard the spacecraft. The transmission is 1200-baud BPSK packet at 145.935 MHz. To make reception as easy as possible for teachers, AMSAT-UK sells a plug-in "dongle" receiver for about $155 at **www.funcubedongle.com**.

As usual, best results are always obtained with directional gain antennas, but teachers have also reported good results using omnidirectional antennas combined with low-noise receive preamplifiers.

The dongle connects directly to any Windows computer on which teachers can run the free "dashboard" program provided by the FUNcube project. The software will decode and display the data captured from the satellite. More than 50 parameters including spacecraft temperature, solar panel voltages, and much more are available for students to enjoy.

For the remaining portion of its orbit, FUNcube-1 operates a linear transponder with an uplink passband from 435.130 to 435.150 MHz, and a downlink passband from 145.950 to 145.970 MHz.

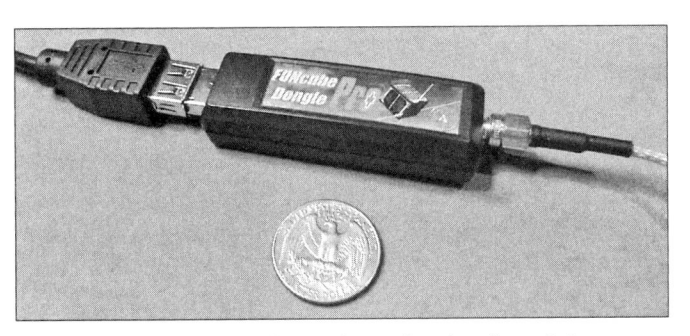

The FUNcube "dongle" receiver simply plugs into a desktop or laptop computer.

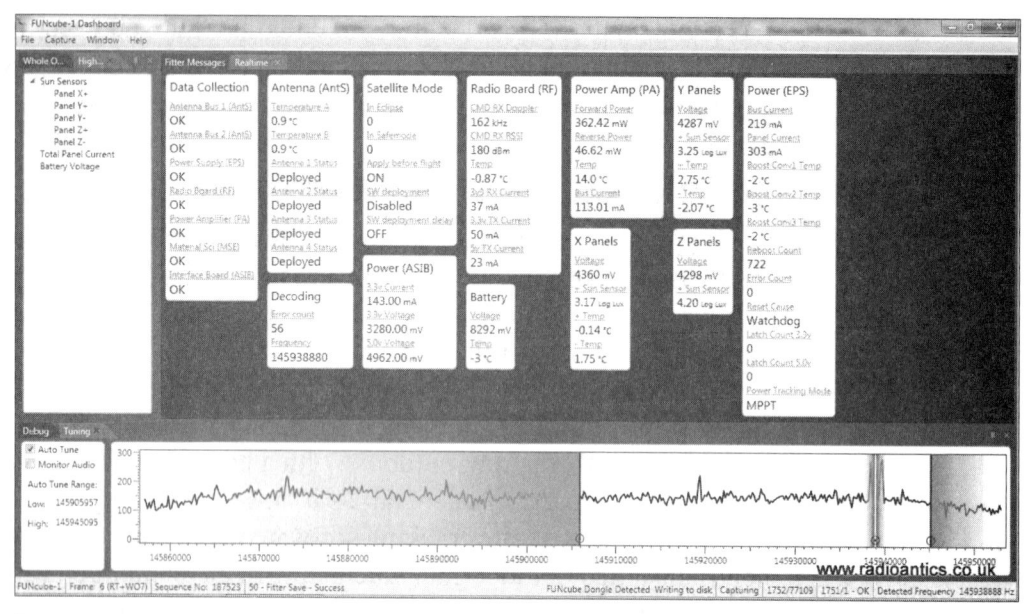

The FUNcube dashboard software allows students to decode and display telemetry from the satellites.

A total of five FUNcube spacecraft have been launched at the time this was written and more are in development. Other than FUNcube-1, the remaining active bird is FUNcube-5, or EO-88. You'll find more information at the FUNcube website at **https://funcube.org.uk/introduction/**.

Research Satellites

Colleges and universities routinely launch CubeSat birds for various research purchases. These small satellites contain scientific experiments, such as measuring cosmic rays, and frequently include amateur radio functionality such as FM repeaters or linear transponders.

Because the satellites are dedicated to research, they send telemetry data that can be received and displayed. In fact, the institutions usually encourage telemetry reception as part of their educational missions.

From a teaching standpoint, the only downside involves the fact that some of the data downlinks use digital communications modes that require equipment (and knowledge) somewhat outside the experience of most amateurs. For example, a research satellite may send its telemetry as a 9600-baud GMSK signal, which would require a radio capable of receiving a 9600-baud signal without distortion and a modem capable of decoding GMSK.

That said, you will still find satellites that use more "ordinary" methods, such as 1200-baud AFSK packet that can be received with a normal FM transceiver and

Universities are often launching CubeSats like this one, and many provide telemetry and can be decoded and displayed.

The Timewave PK96 terminal node controller (TNC) can be used with a compatible transceiver to decode data from satellites.

processed with Terminal Node Controllers (TNCs) such as the Timewave PK96 (**www.timewave.com/product/timewave-pk-96-packet-tnc/**).

Many of these satellites are placed in low-Earth orbits and they have relatively short lifespans. See the current list at **www.amsat.org/beaconclosed-satellites/**. If you're fortunate enough to find a satellite that is nearing reentry, that can be a fascinating exercise for students as they receive the telemetry and observe the changes (such as heating) as the satellite nears its end.

Amateur Radio on the International Space Station

Rosalie White, K1STO
ARISS-US Delegate for ARRL
ARISS-International Secretary-Treasurer

Amateur radio operators are typically curious and like to learn unusual or new things. You're learning about satellites from everything you read in this book. Amateur radio satellites and Amateur Radio on the International Space Station (ARISS) are closely tied together and share many things in common.

You Probably Know a Little About ARISS Already

The ARISS program, Amateur Radio on the International Space Station, exactly matches its name. Special amateur radio rigs set up on the International

The International Space Station as seen from an approaching Russian spacecraft.

Satellites and Education   5-5

Space Station (ISS) let its crew members get on the amateur radio airwaves.

The goals of ARISS are to:

- inspire an interest in science, technology, engineering, and math (STEM) subjects and in STEM careers among young people;
- provide an educational opportunity for students, teachers, and the general public to learn about space exploration, space technologies, and satellite communications;
- provide an educational opportunity for students, teachers, and the general public to learn about wireless technology and radio science through amateur radio;
- provide an opportunity for amateur radio experimentation and evaluation of new technologies;
- provide a contingency communications system for NASA and the ISS crew; and
- provide crew with another means to directly interact with a larger community outside the ISS, including friends and family.

People know ARISS well for its ability to engage young people worldwide with astronauts and the ISS through amateur radio. The ARISS program also aims to draw all ages of people to astronauts, space, and the ISS. ARISS voice contacts can be scheduled for schools and education organizations chosen for their submitted education proposals. Other ARISS contacts are random and you'll read here how hams and non-hams engage with the ARISS ham radio stations using various modes. (For this part of the chapter, schools and education organizations will be referred to as "schools," but please remember that large education organizations have plenty of opportunities for ARISS contacts, just as schools do.)

ARISS scheduled contacts allow young people in schools to experience the excitement of talking directly via amateur radio with crew members on the ISS. An orbiting ISS crew member answers questions from students — kindergarten to post-grads — about science, technology, engineering, math (STEM); radio science, and life in space. The result is often young peoples' increased enthusiasm for these topics, and sometimes a transition into STEM career studies and amateur radio.

How Did Amateur Radio Get on A Space Vehicle in the First Place?

In 1983, 3 days after he launched into space, Astronaut Owen Garriott, W5LFL, talked over the amateur radio airwaves, becoming the first ham radio operator to do so from space. He was on the STS-9 Space Shuttle *Columbia* mission. Owen, who passed away on April 15, 2019, thrilled hams across the globe with first-ever contacts made with a spacecraft orbiting above Earth. Suddenly, hams became the only people to communicate directly with an astronaut in space other than Mission Control and leaders of countries. What an amazing thing, hearing a fellow ham on amateur radio while on a space vehicle! Owen was onboard the shuttle doing research but, in his time off, he used a very special Motorola 2-meter handheld radio, a custom-built window-mounted ham antenna, and a headset. His feat paved the way for all future ham operations on space vehicles. He caused thousands of hams to follow human spaceflight and amateur radio activity in space, and who knows how many people decided to earn their ham license because of it.

ARRL and AMSAT-NA supported Owen's mission and continued to support amateur radio on space vehicles, thanks to NASA, from that day right up to today. The mission proved that ham radio volunteers with special knowledge could put together a space-certified ham radio station and get a space agency's official approval to use it in a space vehicle. Owen captured the excitement of parents and educators, too, who knew kids would listen and be inspired. The Shuttle Amateur Radio EXperiment (SAREX) became a reality.

The next ham operator to fly in space was Astronaut Tony England, WØORE, in 1985 on STS-51, who chose to focus more heavily on schools, making SAREX a true education program. Soon many more astronauts began asking how to get a ham license to talk to and motivate youth. SAREX grew and got to be known around NASA circles as the frequent flyer.

For every crew member who got licensed and was assigned to a shuttle mission, NASA set up the SAREX radio system in the shuttle he or she was launching on, and the system was uninstalled after that shuttle landed back home. SAREX was the forerunner of the ARISS program.

If you want to read more history, go to the ARRL web pages and search on "ARISS history."

Stepping Ahead

NASA knew exactly how successful SAREX had become. In the meantime, the ISS was built and assembled by many people in many nations (and they continue to support it). In 1996, when NASA nearly completed the ISS US laboratory module, the agency told the SAREX team, "If you want amateur radio on the ISS, put together one worldwide team. Lead the project and get your hardware designed, built, and ready for space-certification testing." The tasks: set up a world team, plan the ham station, make it happen, manage the program, and take care of any possible problems that might occur. The Russian Space Agency knew, as well, the significant benefits of having amateur radio onboard *Mir* and the future ISS.

The SAREX team had already considered various hardware ideas, so they jumped on amassing a world team. NASA wanted a group that matched the ISS-supporting space agencies and desired ham leaders from countries that built the ISS. ARRL led the charge in inviting participation by those countries' International Amateur Radio Union (IARU) national ham radio societies (just as ARRL is in the US). AMSAT-NA (Radio Amateur Satellite Corporation) did the same in inviting those countries' national AMSAT groups. The world team chose the name Amateur Radio on the International Space Station, ARISS for short.

ARISS continues to be managed by the international consortium of those ham radio organizations (ARRL and AMSAT-NA in the US) and the related space agencies. The latter include National Aeronautics and Space Administration (NASA), Canadian Space Agency (CSA), Japan Aeronautics Exploration Agency (JAXA), Rosaviakosmos (Roscosmos) in Russia, and European Space

Ready, set, aim: Galileo STEM Academy Ham Radio Club members learn to handle antennas they built themselves, to make satellite contacts.

Agency (ESA). ARISS-International, a working group matched to those five space agencies, consists of five official ARISS Regions: ARISS-Canada, ARISS-Europe, ARISS-Japan, ARISS-Russia, and ARISS-US. Because parts of the world do not quite geographically align to one of the five ARISS regions, they handle matters based on IARU Regions 1, 2, and 3 territories.

Each ARISS region has two delegates — one represents the IARU society and one, the AMSAT group. Delegates work with hundreds of team members (all volunteers) and coordinate with the respective space agencies. ARISS-Europe gets four delegates because of Europe's many countries and regional space organizations tied to ESA. Three officers make up an ARISS-International Board: Chair, Vice Chair, and Secretary/Treasurer. (The author of this part of the book is currently the ARISS-International Secretary-Treasurer and the ARISS-US Delegate for ARRL.) ARISS-International runs the program through monthly teleconference meetings, an annual in-person meeting, and a plethora of emails.

You can read more about all aspects of ARISS by going to the ARISS website, **www.ariss.org**, and going to the ARRL web pages and searching on "ARISS International Partners."

ARISS and Its NASA Sponsors

In the US, currently ARISS's primary sponsors are NASA's Space Communication and Navigation (SCaN), and the ISS National Lab[1]-Space Station Explorers (INL-SSE). SSE is a consortium of organizations that use the power and appeal of the ISS to engage students through educational programs that feature STEM concepts and skills. SCaN and INL-SSE provide funding for the cost of operations support for ARISS at NASA Johnson Space Center in Houston, Texas. This permits ARISS to schedule educational radio contacts, to do ISS crew training in operating the onboard ARISS radios and learning ham radio procedures, and to guide crew about ham radio licensing.

---

[1]About the International Space Station (ISS) US National Laboratory: In 2005, Congress designated the US portion of the ISS as the nation's newest national laboratory to optimize its use for improving quality of life on Earth, promoting collaboration among diverse users, and advancing STEM education. The ISS National Laboratory (INL) manages a portfolio of both basic and applied research projects to support the transition of low Earth orbit to a robust space economy. A fundamental component of the INL mission is to make the unique properties of the low Earth orbit environment available for use to non-NASA US government agencies, academic institutions, and the private sector. Through management of a comprehensive and diverse research portfolio, the INL is driving innovative science that can benefit life on Earth and demonstrate the value of space-based research to the American public.

NASA believes ARISS contributes to STEM learning. At the 2019 ARRL New England Division Convention in Boxboro, Massachusetts, Senior Education Manager Dan Barstow of the ISS National Lab, said in his keynote banquet speech, ARISS is "the power of combining ham radio and space exploration into a magical elixir to engage students."

Educators know about the magic, too. A few months ago, one wrote:

"The students here cannot stop talking about the experience of actually speaking with an ISS crew member via ARISS. Our faculty has now integrated several aspects of amateur radio into the school's STEM curriculum and students formed a school amateur radio club."

The ARISS Ham Radio Stations on the ISS

The very first payload to fly on the ISS, and of course, to receive NASA and Russian certification after many tests, was ARISS! The ARISS hardware team planned all ARISS ham radio stations of the past, the ham stations currently in use, and the newest InterOperable Radio System, which as of late 2019, is nearly ready to be launched to the ISS. For early ISS radios and some still onboard, the team determined what units could be modified to pass NASA's stringent requirements. The ARISS experts also designed, developed, and fabricated totally custom modules for the ARISS ham station. Team members have traveled many times to NASA Johnson Space Center to perform, in concert with NASA certification and test engineers, all of the heavy-duty tests required for any hardware going to the ISS.

The Radios Early On

The early ham stations on the ISS lived in two of the ISS modules. Two weeks after the ISS saw its first inhabitants in November 2000, the very first ISS commander William (Bill) Shepherd, KD5GSL, set up in the Columbus module the Ericsson handheld radios (M-PA commercial series) and a custom packet module, along with supporting cables, adapters, and a headset. The handhelds put out up to 5 W for FM voice and packet — one for the 144 to 146 MHz band and one for the 435 to 438 MHz band. Bill made his first contact merely a few weeks later with the first ARISS-scheduled school, Burbank School in Burbank, Illinois. In December 2003, cosmonauts welcomed a JVC-Kenwood TM-D700 transceiver launched to the Russian Service Module, and a few years later, the ARISS custom SSTV module.

After over a dozen years of thorough use, the 2-meter Ericsson's aging parts resulted in intermittent behavior. Most schools did not have access to 70-centimeter radio equipment, and the average ham found it a bit difficult to deal with UHF Doppler when trying to make contacts, so the crew stowed the Ericsson UHF radio. In 2014, the Russian Energia company provided a JVC-Kenwood TM-D710E transceiver for the Russian Service Module and the crew stopped using the D700 for standard operations.

The Main ARISS Radios in 2019

The JVC-Kenwood TM-D710E supports FM and packet and operates on 144-146 MHz and 435-438 MHz. This radio is restricted to 25 W output power when in operation on the ISS; even so, it provides nearly horizon-to-horizon signal reception for earthbound hams using handheld radios or scanners.

The ARISS team programmed channels in the radio for 2-meter voice as follows:
- 145.80 down/144.49 up for ITU Regions 2 and 3 (the Americas and Asia), and
- 145.80 down/145.20 up for ITU Region 1.

Having two uplink frequencies means that operations follow the worldwide IARU band plan. The crew switches between frequencies; for example, when a crew member begins a contact over the US going east, she or he can track US stations until hitting the Atlantic and quickly loses US stations. Then the crew member switches to the other frequency and picks up stations in Europe or Africa.

The ARISS Antennas

The Columbus module has five ham antennas that support 2 meters, 70 centimeters, L-band and S-band, and serve the HamTV DATV transmitter (more on this later). The ISS crew operates the 2-meter packet radio in unattended mode, allowing hams to make contacts to the ISS ham radio station when crew members are busy working. Hams can communicate with each other using ARISS's packet radio mode or receive SSTV images.

The Russian Service Module offers four antenna systems for the current Kenwood D710E radios. Astronauts and cosmonauts installed and deployed these antennas during various spacewalks in 2002. Each antenna can support ham operations on multiple frequencies and allow simultaneous automatic and crew-tended operations. Having four antennas ensures ARISS operations continue even if an antenna fails. Three of the four antennas are identical and work on 2 meters, 70 centimeters, L-band and S-band transmit, and receive operations. The fourth antenna is a 2.5-meter-long vertical whip that could handle HF operations, particularly 10 meters, but the ARISS team's funding (at the time this book was written) doesn't cover the development of 10-meter space-certified equipment.

A Brand New Sophisticated ARISS Ham Radio Station

The ARISS team knew radios would not last in space for years with cosmic radiation possibilities and wear and tear. Years back, they set to work planning a new ham station. They named it the InterOperable Radio System (IORS), for its ability to work in any of the ISS modules with their differing voltages, thus interoperable all over the ISS. The new radio system consists of a JVC-Kenwood D-710GA transceiver with specially set up software and the Multi-Voltage Power Supply, a custom-built unit developed, designed, and fabricated by ARISS volunteers on the ARISS hardware team. The explanation for the IORS and the team's great progress over time on preparing it, along with NASA's four certification phases were explained in depth by Joe Lynch, N6CL, in the July 2018 issue of ARRL's *QST*.

On different weeks in July and August 2019, the IORS successfully completed a battery of space-certification tests. RF tests for electromagnetic interference (EMI) and electromagnetic compatibility (EMC) verified ARISS hardware would not interfere with the ISS systems or other payloads. Other tests ensured the team built the IORS to operate with the three different voltages of the ISS power sources. The IORS effectively passed power quality tests verifying it would not introduce harmful signals back into the ISS power system. Acoustics tests compiled data on sound pressure levels — first tests were performed at the highest levels of operation so as to understand the acoustic limits, and re-tests were done using levels expected for nominal operation.

ARISS hardware team members Lou McFadin, W5DID, and Kerry Banke, N6IZW, had traveled to the NASA Johnson Space Center with the IORS to support those many days of testing, in concert with the NASA test and certification team.

Kerry explained, "Since the IORS is being qualified to operate on 120 V dc, 28 V dc and Russian 28 V dc (Russia's voltage range to be tested differs from the US voltage range), as well as transmitting on VHF or UHF, a lot of test combinations were required to cover all cases. Each input voltage type was tested at low, medium, and high line voltages. Moreover, additional permutations were required to test the IORS under no load, medium load, and full load at each voltage level. It should not be surprising why the tests took weeks to complete."

Successful completion of each phase of testing meant a milestone in preparing the IORS for launch. In September 2019, the ARISS team began final assembly of the multiple Kenwoods, MVPSs, cabling, and connectors to go to the Columbus module and the Russian Service Module.

In October 2019, the ARISS team proudly announced the hardware team had assembled the first *actual* flight model and it was operating. Kerry put the unit on his sophisticated home test bench, taking it through the paces of quality and verification tests just like NASA's. The team scheduled the very last series of NASA-certification tests at Johnson Space Center for November 2019 to verify all worked within NASA and international space agency requirements and to earn final flight certification.

The team targeted a Spring 2020 liftoff, which meant the entire new ARISS radio system with cabling and connectors had to be delivered to NASA in December 2019. Additionally, in December 2019, the team completed the fourth and last of four NASA safety reviews. Naturally, continuing to update you, the reader, is impossible once this book is printed. The best way to keep apprised of the new radio system is to go to **www.ariss.org**.

The Average Ham Radio Operator Can Contact the ARISS

Astronauts have long workdays, and during their time off, pre-sleep time, and before and after mealtime, enjoy various activities — playing music, photographing our beautiful blue planet, and some crew members who've earned their ham license enjoy getting on the air for fun. They've made voice contacts with thousands of hams around the world. One ham who made an ARISS contact a few months back, wrote: "It was certainly a thrill hearing my call sign come back from the ISS!"

As described in previous chapters, hams on the ground need basic amateur radio equipment to contact astronauts on the ISS. Modes include voice, packet APRS, and SSTV. Some hams will be involved with ARISS HamTV again, too. Hams find published orbital schedules for when the ISS will be overhead and follow radio practices like those described in this book for satellite operations.

Hams with a typical ground station for contacting the ARISS ham station use a 2-meter FM transceiver and 25 – 100 W of output power. A great type of antenna is a circularly-polarized crossed-Yagi antenna that you point in both azimuth (North-South-East-West) and elevation (degrees above the horizon). Many hams, though, have made successful contacts with a vertical or ground plane antennas.

ISS crew members use these call signs from the ARISS radio stations:
- Russian: RSØISS
- USA: NA1SS
- European: DPØISS, OR4ISS, IRØISS
- Packet Station Mailbox: RSØISS-11 and RSØISS-1

Other call signs may be used as the station and crew change. Just as with amateur satellite operations, most ARISS operations are split-frequency — each ham station uses separate receive and transmit frequencies. Amateurs listen to the downlink frequency when the ISS is in range of their location. They transmit on the uplink frequency when crew members might be on the radio. They *do not* transmit on the ARISS downlink frequency.
- Voice and SSTV Downlink: 145.80 MHz (Worldwide)
- Voice Uplink: 144.49 MHz for ITU Regions 2 and 3 (the Americas, and the Pacific and Southern Asia)
- Voice Uplink: 145.20 MHz for ITU Region 1 (Europe, Russia, and Africa)
- VHF Packet Uplink and Downlink: 145.825 MHz (Worldwide)
- UHF Packet Uplink and Downlink: 437.550 MHz (rarely operational)
- UHF/VHF Repeater Uplink: 145.99 MHz (CTCSS 67 Hz)
- UHF/VHF Repeater Downlink: 437.80 MHz

The crew turns off ARISS radios during spacewalks and vehicle dockings and undockings per NASA's rule.

Voice

The crew's usual awake period is 0730 to 1930 UTC. Typical crew leisure times are about one hour after waking and before sleeping, and at times on the weekend. To get a better idea of an astronaut's day, do an online search for "an astronaut's work" to see URLs of NASA web sources on crew schedules. A ham who snagged an ARISS contact wrote, "I am super proud that the ISS crew takes time for us ham radio operators!"

Packet and Digipeating

The Columbus module contains the Ericsson handheld transceiver for operation as a packet digipeater. It uses the call sign RSØISS and responds to the alias "ARISS."

However, the ARISS packet module is old. Sometimes it works, sometimes it doesn't. The "fix" will be the launch of the new ARISS radio system, the IORS that you read about earlier.

To watch for updates, go to **www.ariss.org** and look for the tabs at the top of the page, and hover over "General Contacts" to spot the dropdown box that lists "Packet/APRS" as one choice. Updates on digipeating were at **www.ariss.net**; the site listing current and past digipeating activity and other information.

SSTV

More and more people love taking part in the ARISS SSTV sessions. An ARISS team member's friend, who is a longtime DXer, wrote about his ARISS SSTV experience:

"Picture an adult man in his suburban front yard Saturday morning, in his sweatshirt, slippers, and pajama bottoms pointing a Yaesu handheld radio in the air and recording a sound stream into his smartphone — with little idea of what he was doing or what would happen (and hoping the neighbors were not watching). After messing around through trial and error, I obtained an image playing the recorded audio into the SSTV app on my IPad. Maybe not a great print, but one of the coolest things I've done in amateur radio!"

The crew utilizes the Kenwood D710E to transmit JPEG still images that earthbound hams, space fans, and shortwave listeners have fun downloading. To transmit the specially-stored images, the onboard SSTV equipment employs *MMSSTV* software (mentioned in a previous chapter), a radio/computer interface module, and data cables.

The ARISS transmissions are made on 145.800 MHz FM in the SSTV mode

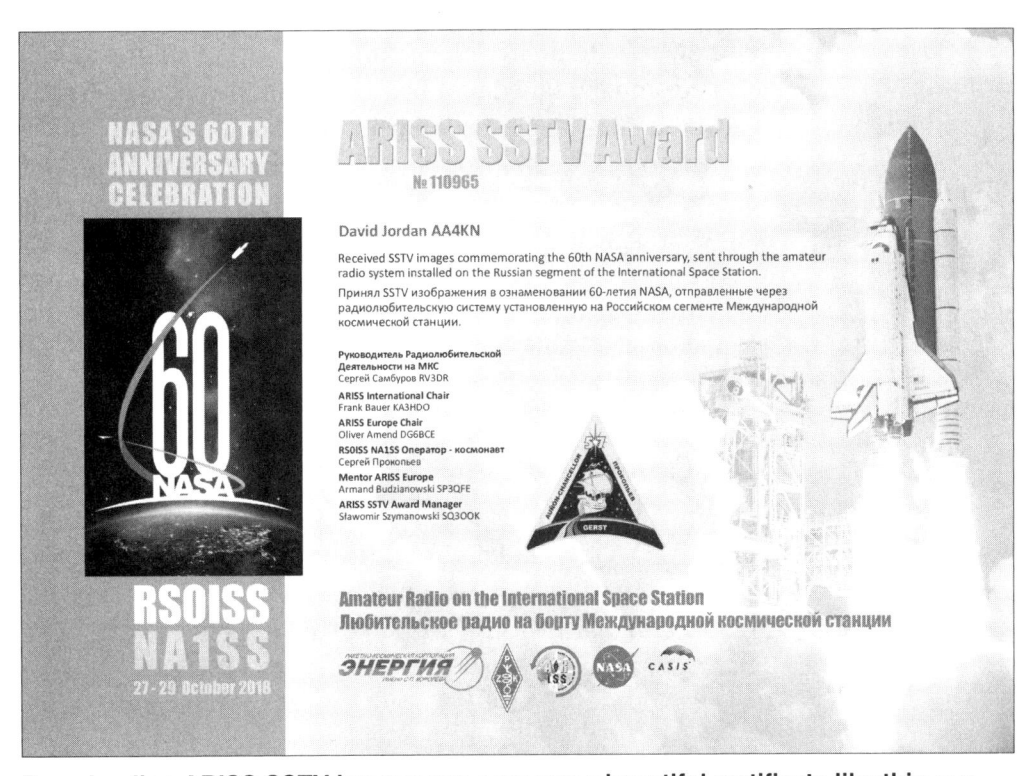

Downloading ARISS SSTV images can earn you a beautiful certificate like this one.

PD-120. Hams show off their downloaded images on this ARISS SSTV Gallery at **www.spaceflightsoftware.com/ARISS_SSTV/index.php**. You can learn more from FAQs at **www.spaceflightsoftware.com/ARISS_SSTV/faq.php**.

Hams can receive a special handsome ARISS SSTV award for posting their images; learn more at **https://ariss.pzk.org.pl/sstv**.

The ARISS team distributes the latest news for SSTV transmission sessions on ARISS Facebook and on ARISS Twitter @**ARISSstatus**.

You can try several methods to download ARISS SSTV images. John Brier, KG4AKV, an avid ARISS supporter, created a video demonstration of receiving SSTV images from the ISS. He prepared an excellent online tutorial as well, explaining in more detail how to configure a simple system to acquire and view ARISS SSTV images. Links to both are: **www.youtube.com/watch?v=7to9uX1sWC4** and **https://spacecomms.wordpress.com/iss-sstv-reception-hints/**.

HamTV

On February 11, 2016, the first ARISS HamTV transmission thrilled hundreds of people during an ARISS school radio contact at the Royal Masonic School for Girls in Rickmansworth, UK. The normal audio-only ARISS school contact was enhanced with video! Astronaut Tim Peake, KG5BVI/GB1SS, activated the Ham Video transmitter on board Columbus for the first time, and everyone in range on the ground could watch the astronaut who was talking to them via the ARISS radios. The HamTV unit had been launched and set up on the ISS long before, followed by a great deal of testing prior to its first success at the UK school.

Alas, in April 2018, the Ham Video unit stopped working. The ARISS team expended a lot of time and work on studying the problem, acquiring approval for crew time to do troubleshooting onboard, and in getting NASA to package and ship the failed unit back to Earth. It rode on a SpaceX vehicle and was recovered after it splashed down in the Pacific Ocean. The team had the unit shipped to NASA Johnson Space Center and on to Italy, where the video module had been developed and custom built.

After a great deal of testing by the manufacturer, engineers determined in October 2019 that a radiation hit caused the problem. A team performing repairs hopes that all the steps needed, such as testing, paperwork, shipment back to the US, and so on, will result in re-launching the repaired Ham Video unit in mid-2020.

QSL Cards

When you connect with an ARISS onboard radio station you can get a coveted ARISS QSL card confirming contact. To learn how, go to **www.ariss.org/qsl-cards.html**. A ham who made an ARISS contact and recently requested an ARISS QSL card wrote: "My first ARISS contact! This was a real treat for me; what a great hobby. I can't wait to get my QSL card!"

A packed gym at Hudson (New Hampshire) Memorial School aided by Nashua Area Radio Society.

ARISS School Contacts

If you are an educator, or know someone who is, you should know that one of the primary ARISS missions is to provide crewmember contacts to educate youth.

An ARISS radio contact scheduled for a school is a real-time voice-only ham contact with an ISS crewmember. As the ISS moves from horizon to horizon over a school, the length of the radio contact will be about 10 minutes long. During the contact, young people interact in a question-and-answer format. During ARISS contacts, audiences learn firsthand from astronauts about working, living, and communicating in space.

ARISS officially invites schools (formal and informal education organiza-

Young people do get excited hearing the astronaut via ham radio — even the boy who can't believe what he's hearing!

Satellites and Education   5-15

tions and groups), individually or working together, to develop an education proposal to submit to ARISS. To get great outcomes from the 10-minute ISS pass overhead, the ARISS program looks for schools that can draw large numbers of participants (such as an entire school and many area citizens), make the radio contact part of a well-developed education plan, and compile a media plan. A school's proposal must show it can be very flexible if having to accommodate a change of a scheduled ARISS contact. Many reasons can make it necessary to change the date and time of a scheduled ARISS contact — delayed vehicle launches due to weather, their subsequent dockings and undockings (remember, the ARISS radios get shut off at those times), switches in crewmembers' sleep periods, and on and on.

Every 6 months, ARISS-US and ARISS-Europe open a window for accepting ARISS Education Proposals from schools. The ARISS-US team distributes a news release to ARRL, AMSAT, NASA, and many ham radio news outlets to announce the opening of a window. Of course, the ARISS team posts the details at **www.ariss.org**.

You or your favorite educator will see the proposal information, expectations, the *ARISS Proposal Guide*, and *ARISS Proposal Form* on the **www.ariss.org** website and searching on: "Hosting an ARISS Contact in the USA." Questions about ARISS go to **ariss.us.education@gmail.com**. As a proposal window opens, the ARISS team sets dates and times of ARISS Proposal Webinars and posts that information at **www.ariss.org**.

If you live in a country other than the US, search on: "Applying to Host a Scheduled Contact" and look for the paragraph that covers your part of the world.

After reading the *ARISS Proposal Guide*, the school fills out and submits the *ARISS Proposal Form* for an opportunity of a lifetime. If the proposal rises above the others, the school jumps the first hurdle; then the school must compile a ham radio equipment plan. At that time, ARISS matches an ARISS Technical Mentor — in the US and around the world — to assist every school selected.

The schools look for local hams who can put together, and temporarily set up in a school auditorium or gym, good satellite-tracking ham equipment and have operational expertise to make an ARISS direct radio contact. The equipment must meet certain technical requirements to ensure a successful contact; the ARISS Mentor approves the equipment list. ARISS recommends equipment configurations that are found at **www.ariss.org/submit-a-contact-proposal.html**. Scroll down that web page a few paragraphs to spot the information.

If there is no local amateur

Teacher Kathy Lamont, KM4TAY, got licensed after submitting her ARISS Education Proposal. Here she demonstrates satellite operations to her school's students. Her local ham club tasks her each year at Field Day with making the satellite contact for the bonus point.

After being selected for an ARISS contact, these teachers applied for ARRL Teachers Institute (TI). (For details, browse on Teachers Institute at ARRL's web pages). They studied for their ham licenses before going to TI and while there, passed their exams. Beth Bivens (left) is now KB7SKI and Gina Kwid is KJ7ICO.

radio club, the ARISS Technical Mentor advises on other options and equipment needed for each situation. Keep reading for more about those.

First, you may ask: Are today's young people and today's ham operators even interested in space? Here is a quote from a university student: "Thank you very much for the ARISS contact! You have inspired our students into a new generation of scientists and engineers." Another student wrote: "I am a brand-new amateur radio operator and I'm 14 years old. This packet contact is going to get me my first QSL card!"

What if a School Can't Assemble the Required Satellite Station?

If an amateur club cannot assist in setting up and operating a satellite station for the contact, the solution may be a *telebridge* radio contact.

A telebridge contact utilizes an amateur with a suitable station, who may actually be quite far away from the school. This volunteer communicates with the ISS's ham station and simultaneously relays the audio to and from the school.

Both an ARISS direct radio contact and an ARISS telebridge radio contact give young people the chance to speak via amateur radio directly to an astronaut and learn how a ham radio station works. The area club offers students and educators (some hams, some not) hands-on ham radio activities. Fox hunts, getting youth on the air, helping them build simple handheld antennas or electronic kits, downloading ARISS SSTV images, viewing NOAA weather images, trying satellite contacts, and whatever ham activities club members enjoy.

Each selected school is assigned an ARISS Education Ambassador along with the ARISS Technical Mentor. An ambassador helps the school's educators stay on track with their STEM lesson plans and education activities, including ham lessons, the media plan, and more. Yes, hams and educators do a great deal to make a scheduled school ARISS contact a success, but helpers abound.

The Future of ARISS Depends on Your Support

The ARISS-International team donates approximately $5 million per year of in-kind support to the ISS program, primarily through technical and educational volunteer assistance to schools, space flight radio hardware development, and ARISS operations.

The ARISS team includes all types of volunteers, ham operators like you and me; ARISS Technical Mentors, scheduling/technical representatives, an orbital prediction specialist, ARISS Educator Ambassadors, hams working on hardware and safety documentation, helping with fund-raising and public relations, doing updates to **www.ariss.org**. An ARISS operations team meets weekly by teleconference and exchanges daily emails. ARISS Technical Mentors (experienced ham satellite operators) work with the schools, teachers, and local amateur radio groups that facilitate the radio contacts with the ISS. Scheduling/technical representatives work within the space agencies, primarily NASA in the US and Roscosmos in Russia, to secure actual dates and times for the ARISS radio contacts. Interested? ARISS needs you! Email **ariss.us.education@gmail.com**.

ARISS, just as with any volunteer organization, always needs donations. These go through the Radio Amateur Satellite Corporation, AMSAT-NA, where there is a special fund just for ARISS. If you wish to contribute, you can do so online at **www.amsat.org/donations/ariss-donations/**.

All these donations channel directly to ARISS. ARRL's ARISS web page for donors is **www.arrl.org/amateur-radio-on-the-international-space-station**, which leads to the **www.ariss.org** pages that link to the AMSAT-ARISS donation web page. Donors will receive a heartfelt thank you letter and some ARISS facts.

Contributors who donate $100 or more ($110 for non-US folks) earn the handsome ARISS Challenge Coin.

ARISS's Annual Fund for Operations gratefully accepts donations for continuing its day-to-day program for: (1) operations and training and (2) technical and administration. This includes enhancing educational outcomes, increasing public outreach, improving communications, evolving business development, and compiling and analyzing metrics on outcomes.

ARISS's Next Generation Hardware needs contributions for the cost of upcoming space-certified ham radio systems, future enhancements to the onboard ham radio systems, and at times, repairs to them. For enhancements, costs mount up for development and fabrication, space-rated parts and hardware for necessary custom-built units, hardware space-certification and testing, and cabling and connectors. Because of the dire need for safety, nothing about space hardware is cheap.

ARISS also seeks corporate donations, foundation grants, and organization sponsorships. Interested people can send an email to Rosalie White, K1STO, at **k1sto@arrl.org**.

The Future of ARISS

The International Space Station cannot last forever. It is unclear what year funding will end for habitation of the ISS because of upgrade expenses. NASA and the world's space agencies began looking into the future a few years ago.

They suggested that ARISS International should begin considering Deep Space Gateway and Lunar Gateway. These two ambitious projects will take humanity well beyond Earth orbit.

The ARISS-International team collected some early ideas and pulled together everyone with an interest in the future, eventually naming itself Amateur Radio Exploration or AReX. ARISS-International Chair Frank Bauer,

KA3HDO, and ARISS-International Vice Chair Oliver Amend, DG6BCE, lead the charge. The AReX team meets twice a month and moves forward with many discussions on the Gateways and other Lunar opportunities.

The team began moving in earnest at the 2019 annual ARISS-International Face-to-face Meeting in Montreal, Canada on the two avenues: Deep Space Gateway opportunities and Lunar vicinity activities. Some parts of what *might* be used have already been built by or theorized by ARISS and the AMSAT international communities, such as ARISS HamTV. AReX began considering what's feasible with AMSAT communities — some possible CubeSat projects, where experimentation with these might help lead to information useful for the Gateways.

As ARISS conceptualizes a potential ham radio station on Gateways or for the Lunar vicinity, it will be watchful of, and listening to NASA and other space agencies as their Gateway plans evolve. ARISS heard that the highest probability of getting a radio system on board is to build the hardware to make it available as soon as possible. This was not an official offer from any space agency for ARISS to be part of Gateway or Lunar vicinity activity, but it means ARISS must make a higher priority of building hardware sooner. Stay tuned for more news from ARISS!

Chapter 6
Antenna System Projects

The following projects were originally published in *QST* magazine. Not only are these projects informative and fun to build, you'll save a bundle of money in the process!

By L. B. Cebik, W4RNL

A Simple Fixed Antenna for VHF/UHF Satellite Work

Explore the low-Earth orbiting amateur satellites with this effective antenna system.

When we are just getting interested in amateur satellite operation, the thought of investing in a complex azimuth-elevation rotator system to track satellites across the sky can stop us in our tracks. For starters, we need a simple, reliable, fixed antenna—or set of antennas—to see if we really want to pursue this aspect of Amateur Radio to its limit. We'll look at the basics of fixed antenna satellite work and develop a simple antenna system suited for the home workshop. There will be versions for both 145 and 435 MHz.

Turnstiles and Satellites

For more than decades, many fixed-position satellite antennas for VHF and UHF have used a version of the turnstile. The word "turnstile" actually refers to two different ideas. One is a particular antenna: two crossed dipoles fed 90° out of phase. The other is the principle of obtaining omnidirectional patterns by phasing almost any crossed antennas 90° out of phase. The first idea limits us to a single antenna. The second idea opens the door to adapting many possible antennas to omnidirectional work.

Figure 1 shows one general method of obtaining the 90° phase shift that we need for omnidirectional patterns. Note that the coax center conductor connects to only one of the two crossed elements. A ¼-λ section of transmission line that has the same characteristic impedance as the natural feed point impedance of the first antenna element alone connects one element to the next. The opposing ends of the two elements go to the braid at each end of the transmission line. If the elements happen to be dipoles, then a 70 to 75-Ω transmission line is ideal for the phasing line. However, the resulting impedance at the overall antenna feed point will be exactly half the impedance of one element alone. So we will obtain an impedance of about 35 Ω. For the dipole-based turnstile antenna, we'll either have to accept an SWR of about 1.4:1 or we'll have to use a matching section to bring the antenna to 50 Ω. A parallel set of RG-63 ¼-λ lines will yield about 43 Ω impedance, about right to bring the 35-Ω antenna impedance to 50 Ω for the main coax feed line. For all such systems, we must remember to account for the velocity factor of the transmission line, which will yield a line length that is shorter than a true quarter wavelength.

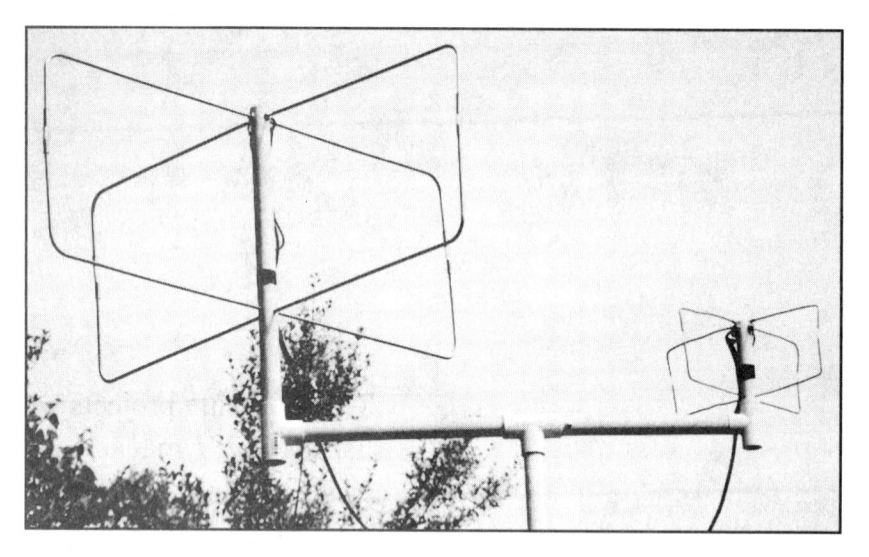

The dipole-based turnstile is popular for fixed-position satellite work. Figure 2 shows—on the left—one recommended system that has been in *The ARRL Antenna Book* since the 1970s. For 2 meters, a standard dipole-turnstile sits over a large screen that simulates ground. Spacing the elements from the screen by between ¼ and ⅜ of a wavelength is recommended for the best pattern. For satellite operation, the object is to obtain as close to a dome-like pattern overhead as possible. The most desirable condition is to have the dome extend as far down toward the horizon as possible to let us communicate with satellites as long as possible during a pass.

The turnstile-and-screen system, while simple, is fairly bulky and prone to wind damage. However, the turnstile loses performance if we omit the screen. One way to reduce the bulk of our antenna is to find an antenna with its own reflector. However, it must have a good pattern for the desired goal of a transmitting and receiving dome in the sky. The dual Moxon rectangle array, shown in outline form on the right of Figure 2, offers some advantages over the traditional turnstile. First, it yields a somewhat better dome-like pattern. Second, it is relatively easy to build and compact to install.

Almost every fixed satellite antenna shows deep nulls at lower angles, and the number of nulls increases as we raise the antenna too high, thus defeating the desire for communications when satellites are at low angles. Figure 3 shows the elevation

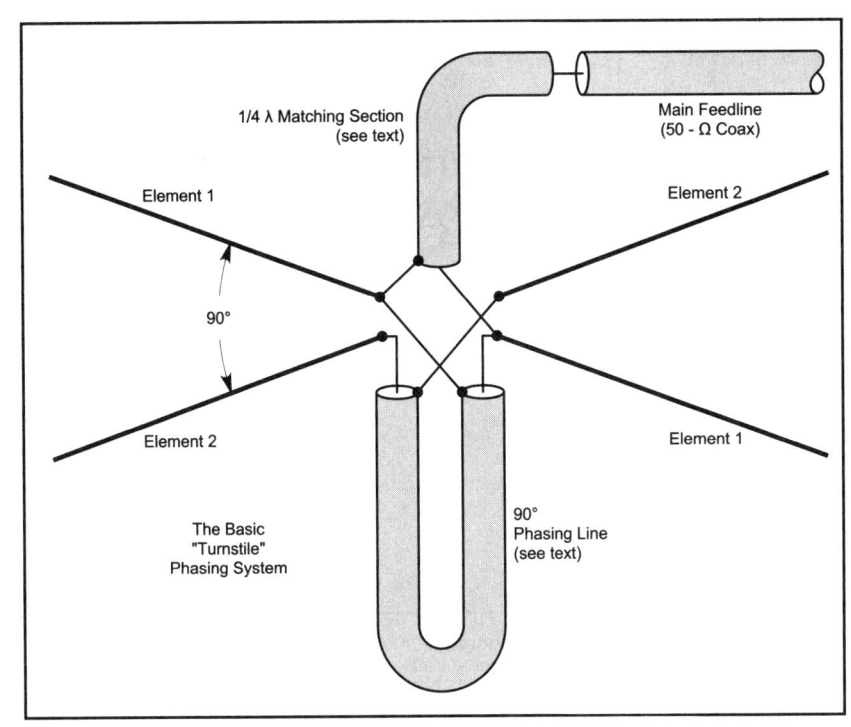

Figure 1—The basic turnstile phasing (and matching) system for any antenna set requiring a 90° phase shift between driven elements in proximity.

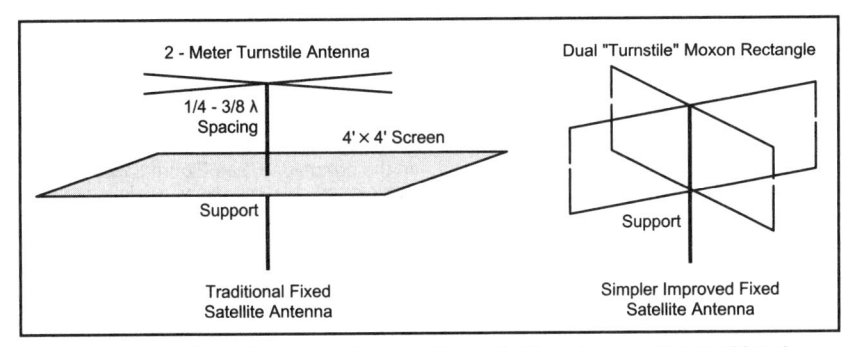

Figure 2—Alternative schemes for fixed-position satellite antennas: the traditional turnstile-and-screen and a pair of "turnstiled" Moxon rectangles.

patterns of a turnstile-and-screen and of a pair of Moxon rectangles when both are 2λ above the ground. A 1λ height will reduce the low angle ripples even more, if that height is feasible. However, the builder always has to balance the effects of height on the pattern against the effects of ground clutter that may block the horizon.

The elevation patterns show the considerably smoother pattern dome of the Moxon pair over the traditional turnstile. The middle of the turnstile dome has nearly 2 dB less gain than its peaks, while the top valleys are nearly 3 dB lower than the peaks. The peaks and valleys can make the difference between successful communications and broken-up transmissions. So, for the purpose of obtaining a good dome, the Moxon pair may be superior.

A reasonable suggestion offered to me was simply to add reflectors to a standard dipole turnstile and possibly obtain the same freedom from a grid or screen structure. Figure 4 shows the limitation of that solution. The result of placing reflectors behind the dipole turnstile is a pair of crossed 2-element Yagi beams fed 90° out of phase. The pattern is indeed circular and stronger than that of the Moxon pair. However, the beamwidth is reduced to only 56° at the half-power points. The antenna would make an excellent starter for a tracking AZ-EL rotator system, but it does not have the beamwidth for good fixed-position service.

The Moxon pair, with lower but smoother gain across the sky dome, offers the fixed-antenna user the chance to build a successful beginning satellite antenna. The pattern will be circular within under a 0.2-dB difference for 145.5 to 146.5 MHz, and within 0.5 dB for the entire 2-meter band. Since satellite work is concentrated in the 145.8 to 146.0 MHz region, the broadbanded antenna will prove fairly easy to build with success. A 435.6 MHz version, designed to cover the 435 to 436.2 MHz region of satellite activity will have an even larger bandwidth.

Like the dipole-based turnstile, the Moxons will be fed 90° out of phase with a ¼-λ phasing line of 50-Ω coaxial cable. The drivers will be connected just as shown in Figure 1. Since the natural feed-point impedance of a single Moxon rectangle of the design used here is 50 Ω, the pair will show a 25-Ω feed-point impedance. Paralleled ¼-λ sections of 70- to 75-Ω coaxial cable will transform the low impedance to a good match for the main 50-Ω coaxial line to the rig. In short, we have "turnstiled" the Moxon rectangles into a reasonable fixed-position satellite antenna.

Building the Moxon Pairs

The Moxon rectangle is a modification of the reflector-driver Yagi parasitic beam. However, instead of using linear elements, the driver and reflector are bent back toward each other. The coupling between the ends of the elements combined with the coupling between parallel sections of the elements combine to produce a pattern with a broad beamwidth. By carefully selecting the dimensions, we can obtain both good performance (meaning adequate gain and an excellent front-to-back ratio) and a 50-Ω feed point impedance.[1]

In fact, a single Moxon rectangle might be used on each band for reasonably adequate satellite service. When pointed straight up, the Moxon rectangle pattern is a very broad oval, although not a circle. The oval pattern also gives the Moxon another advantage over dipoles in a turnstile configuration. If the phasing-line between dipoles is not accurately cut, the normal turnstile near-circle pattern degrades into an oval fairly quickly because the initial single dipole pattern is a figure **8**. The single Moxon oval pattern allows both dimensional inaccuracies and phasing-line inaccuracies of considerable amounts before degrading from a nearly perfect circle.

Figure 5 shows the critical dimensions for a Moxon rectangle. The lettered references are keys to the dimensions in Table 1. The design frequencies for

---

[1] See "Having a Field Day with the Moxon Rectangle," *QST*, June, 2000, pp 38-42, for further details on the operation of the Moxon rectangle, along with the references in the notes to that article. Also included in the notes is the source for a program to calculate the dimensions for a 50-Ω Moxon rectangle for any HF or VHF frequency using only the design frequency and the element diameter as inputs.

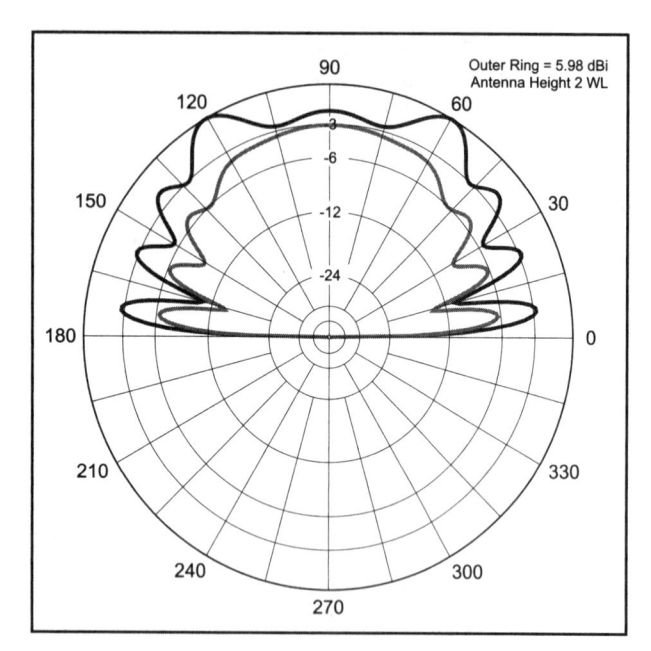

Figure 3—A comparison of elevation patterns for the turnstile-and-screen system (with ⅜λ wavelength spacing, shown in blue) and a Moxon pair (shown in red), both at 2λ height.

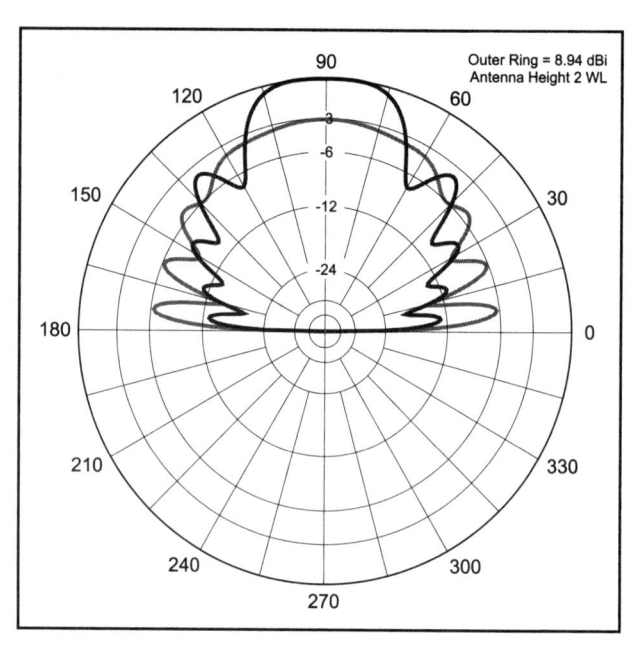

Figure 4—A comparison of elevation patterns for 2-element turnstiles (crossed 2-element Yagis, shown in blue) and a Moxon pair (shown in red), both at 2λ height.

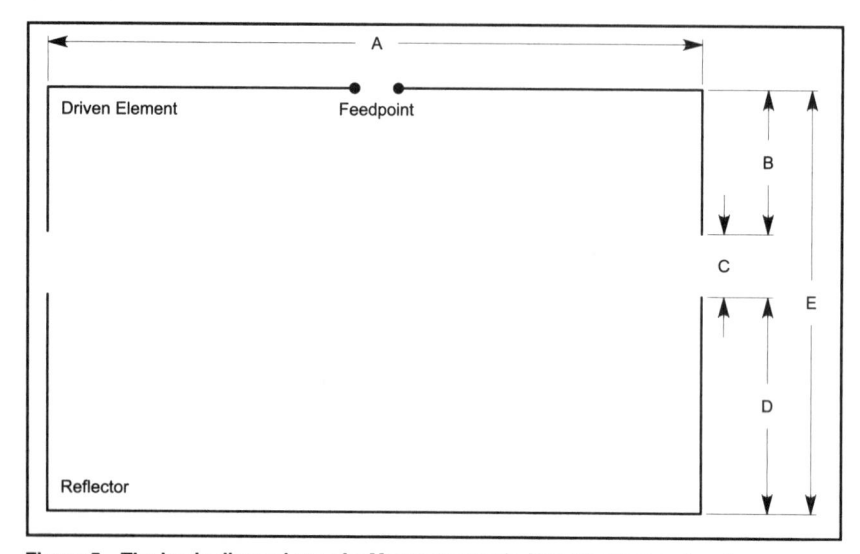

Figure 5—The basic dimensions of a Moxon rectangle. Two identical rectangles are required for each "turnstiled" pair.

Table 1
Dimensions for Moxon Rectangles for Satellite Use

Two are required for each antenna. The phase-line is 50-Ω coaxial cable and the matching line is parallel sections of 75-Ω coaxial cable. Low power cables less than 0.15 inches in outer diameter were used in the prototypes. See Figure 5 for letter references. All dimensions are in inches.

Dimension	145.9 MHz	435.6 MHz
A	29.05	9.72
B	3.81	1.25
C	1.40	0.49
D	5.59	1.88
E (B + C + D)	10.80	3.62
¼ wavelength	20.22	6.77
0.66 velocity factor phasing and matching lines	13.35	4.47

the two satellite antenna pairs are 145.9 MHz and 435.5 MHz, the centers of the satellite activity on these two bands. The 2-meter Moxon prototype uses 3/16-inch diameter rod, while the 435 MHz version uses #12 AWG wire with a nominal 0.0808-inch diameter. (Single Moxons built to these dimensions would cover all of 2-meters and about 12 MHz of the 432 MHz band.) Going one small step up or down in element diameter will still produce a usable antenna, but major diameter changes will require that the dimensions be recalculated.

The reflectors are constructed from a single piece of wire or rod. I use a small tubing bender to create the corners. The rounding of the corners creates a slight excess of wire for the overall dimensions in the table. I normally arrange the curve so that the excess is split between the side-to-side dimension (A) and the reflector tail (D). Practicing on some scrap house wire may make the task go well the first time with the actual aluminum rod. The total reflector length should be A + (2 × D).

The driver consists of two pieces, since we'll split the element at its center for the feeding and phasing system. I usually make the pieces a bit longer before bending and trim them to size afterwards. The total length of the driver, including the open area for connections, should be A + (2 × B).

Perhaps the most critical dimension is the gap, C. I have found nylon tubing, available at hardware depots, to be very good to keep the rod ends aligned and correctly spaced. When everything has been tested and found correct, a little super-glue on the tubing ends and aluminum stands up to a lot of wind. I usually nick the aluminum just a little to let the glue settle in and lock

6-4   Chapter 6

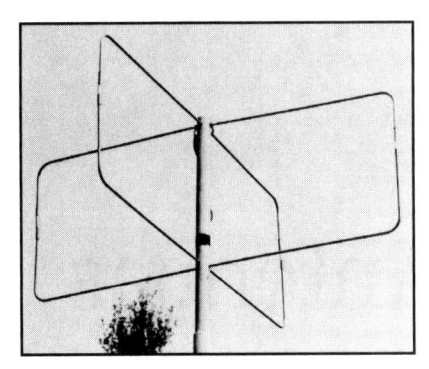

A close-up view of the 145.9 MHz rectangle pair.

the junction. For the UHF version, a short length of heat-shrink tubing provides a lock for the size of the gap and the alignment of the element tails.

It is one thing to make a single Moxon and another to make a working crossed pair. Figure 6 shows the general scheme that I used for the prototypes, using CPVC. (Standard schedule 40 or thinner PVC or fiberglass tubing can also be used.) The support stock is $3/4$ inch nominal. The reflectors go into slots at the bottom of the tube and are locked in two ways. Whether or not the two reflectors make contact at their center points makes no difference to performance, so I ran a very small sheet screw through both 2-meter reflectors to keep their relative positions firm. I soldered the centers of the 435-MHz reflectors. Then I added a coupling to the bottom of the CPVC to support the double reflector assembly and to connect the boom to a support mast. Cementing or pressure fitting the cap is a user option.

The feed point assemblies are attached to solder lugs. The phasing line is routed down one side of the support, while the matching section line is run down the other. Electrical tape holds them in place. For worse weather, the tape may be over-sealed with butylate or other coatings. Likewise, the exposed ends of the coax sections and the contacts themselves should be sealed from the weather. The details can be seen—as built for the experimental prototypes in one of the photos—before sealing, since lumps of butylate or other coatings tend to obscure interesting details.

The overall assembly of the two antennas appears in the second photograph. The PVC from the support Ts can go to a center Tee that also holds the main support for the two antennas. A series of adapters, made from miscellaneous PVC parts to fit over a standard length of TV mast. Alternatively, the antennas can be separately mounted about 10 feet apart. The 10-foot height of the assembly has proven adequate for general satellite reception, although I live almost at the peak of a hill.

The antennas can be mounted on the same mast. However, for similar sky-dome patterns, they should each be the same number of wavelengths above ground. For example, if the 2-meter antenna is about two wavelengths up at about 14 feet or so, then the bottom of the 435-MHz antenna should be only about 4.5 feet above the ground. Placing the higher-frequency antenna below

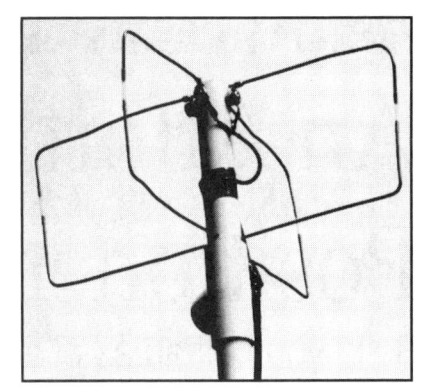

The 435-MHz Moxons.

the 2-meter assembly will create some small irregularities in the desired dome pattern, but not serious enough to affect general operation.

There is no useful adjustment to these antennas except for making the gap between the drivers and reflectors as accurate as possible. Turnstile antennas show a very broad SWR curve. Across 2 meters, for example, the highest SWR is under 1.1:1. However, serious errors in the phasing line length can result in distortions to the desired circular pattern. There is no substitute for checking the lengths of the phasing line and the matching section several times before cutting. The correct length is from one junction to the next, including the portions of exposed cable interior.

These two little antennas will not compete with tracking AZ-EL rotating systems for horizon-to-horizon satellite activity. For satellite work, however, power is not always the problem (except for using too much) and modern receiver front-ends have enough sensitivity to make communication easy. So when the satellite reaches an angle of about 30° above the horizon, these antennas will give a very reasonable account of themselves. When you become so addicted to satellite communication that you invest in the complete tracking system, these antennas can be used as back-ups while parts of the complex system are down for maintenance!

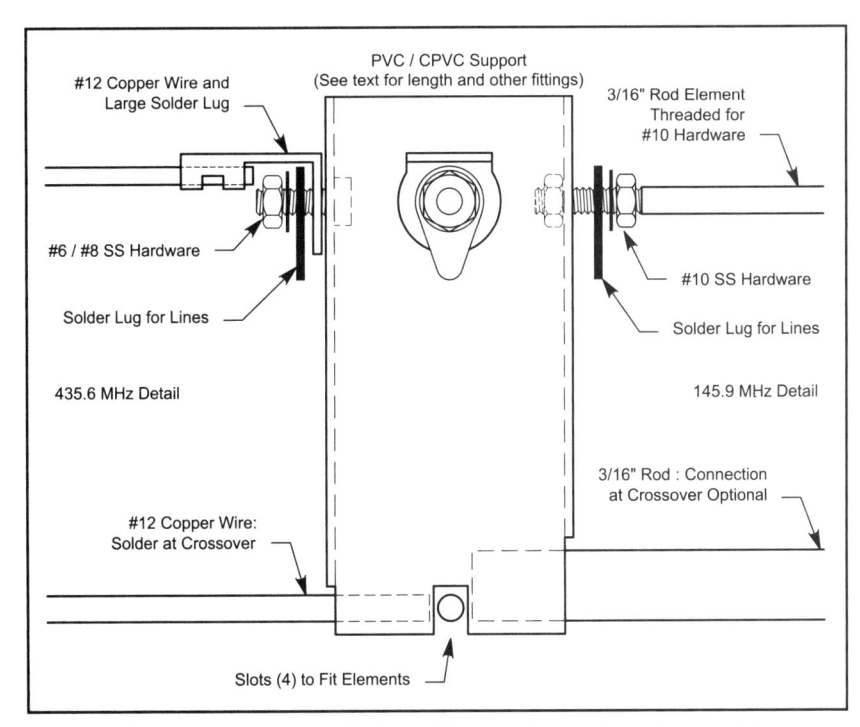

Figure 6—Some construction details for the Moxon pairs constructed as prototypes.

By Lilburn Smith, W5KQJ

An Affordable Az-El Positioner for Small Antennas

Build yourself an az-el antenna positioner for one third the cost of a commercial unit and make it easy to point that antenna.

Why an Az-El Positioner?

Interest in the UHF and microwave bands is at an all-time high. This is probably due to AO-40 and other satellites, and to the increased offering of commercial gear that includes the frequencies through 1296 MHz. In addition, several firms now market transverters for amateur use up to 10 GHz. The surplus market has yielded a number of 10 GHz and 24 GHz components. More technically minded amateurs are experimenting with even higher frequencies.

All the bands above 430 MHz benefit from small, directional antennas. Although it is possible to make microwave contacts without positioning a high-gain antenna accurately, it is too much work to be fun. When you get above 10 GHz, it is almost impossible to make contacts reliably without an antenna positioner.

The AO-40 satellite has two working transmitters: S band (2.4 GHz) and K band (24 GHz). It utilizes two receivers: U (435 MHz) and L (1260 MHz). Although the AO-40 orbit is such that it does not move in azimuth and elevation rapidly, accurate aiming of the antenna is preferred. A small az-el antenna rotator that can handle a UHF or an L band transmit antenna and an S band or K band receive antenna adds to the enjoyment of working the satellite. And, for polar LEO satellites, an az-el positioner is a must.

Why Build Your Own Rotator?

An affordable azimuth antenna rotator is not difficult to find. Cornell-Dubilier Electronics (CDE) and its successors built thousands of the *AR* and *Ham* series rotators. Rugged and reliable, they are practically indestructible, as long as water does not get into the case. Parts are readily available, although expensive. Every hamfest has a few rotators and control boxes. Used, the rotator and box sell for about $100, and a new set of bearings and a new position potentiometer bring the cost up to around $150. The elevation axis is a different story. Although many good rotators were manufactured that allowed the mast to extend all the way through the rotator, they are now practically impossible to find, even used. New az-el antenna rotators are available, but expensive. For example, the Yaesu 5500 is convenient and works well, but will set you back $700.

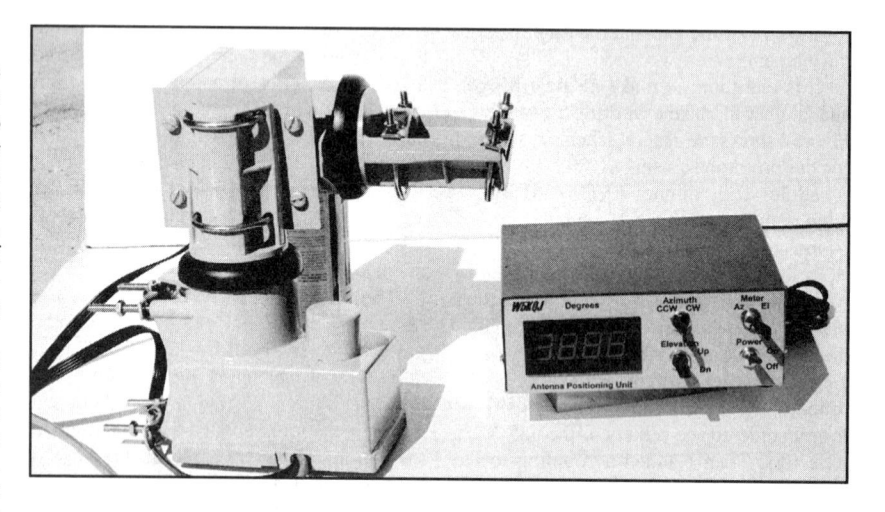

One by one, the former rotator manufacturers ceased to manufacture, until all the inexpensive rotators now being sold new appear to be coming from the same factory. The approach discussed in this article is not new, but the rotators used in past years are no longer available. The RadioShack *Archerotor* and the Channel Master *Colorotor* are actually the same unit. Both models have eliminated the position potentiometer. Also, the rotator does not allow a mast to be run through the unit for the elevation axis. Although there are several ways around the mast problem, the position potentiometer is an absolute must. The rotators sold use a control box with a dial rotated by a motor that runs at the same speed as the rotator. The dial just rotates for the same time interval as the rotator. Synchronization is maintained by occasionally rotating the antenna and dial to their mechanical stops: first clockwise, then counterclockwise. The error builds up until resynchronization is necessary. Although tolerable as a TV antenna rotator, the position-indicating system is unsatisfactory for amateur use. Switching rotation directions rapidly while trying to peak on a satellite or distant station just confuses the indicator.

The latest RadioShack and Channel Master catalogs list only a remotely controlled rotator. Azimuth positions that correspond to TV channel numbers are stored in a microprocessor and are transmitted to the control box by an infrared link. Obviously, this scheme has no application to satellite tracking.

This article will address the mechanical mounting of the antennas in an az-el configuration and the addition of a position feedback potentiometer and a simple control box with a digital readout. Although limited in performance, the whole system can be built for $250 if all new parts are purchased. If flea market parts are used, the whole thing can be built for under $100. The rotators are readily available at hamfests and flea markets because they were so poorly suited to amateur use without the position potentiometer. I bought two brand new ones for $20 each. This az-el positioner will handle any small satellite array and the cost is reasonable.

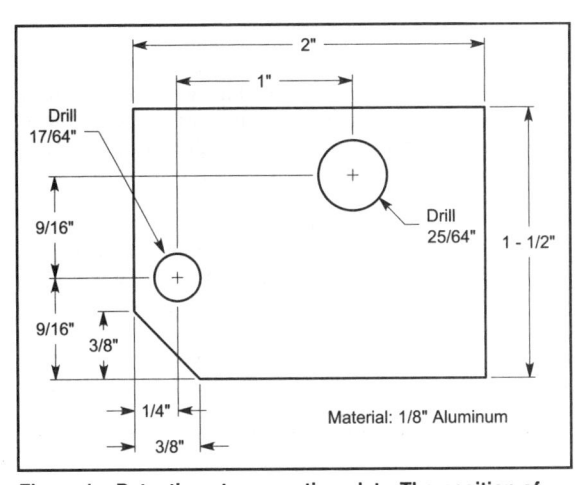

Figure 1—Potentiometer mounting plate. The position of the potentiometer mounting hole is critical for proper gear-meshing. Drill the plate accurately and mark the blank plate before drilling.

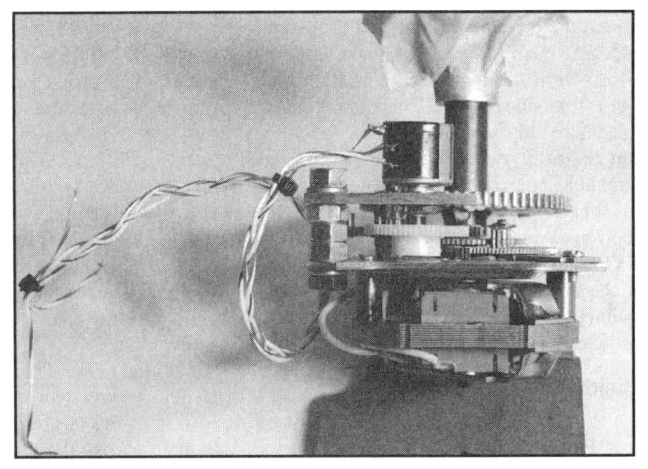

Figure 2—Potentiometer mounting details. Make sure the gears mesh properly, the potentiometer shaft turns freely and the shaft doesn't hit the potentiometer mechanical stops before the rotator shaft stops.

Modifying the Rotator for Positional Feedback

The position feedback potentiometer will be added to make the rotator useful for ham radio. Without it, it has little or no application. Doug Braun, NA1DB, worked out a simple but elegant method to add a 10-turn position pot to the rotator. The potentiometer is easy to add, but it does add about $16 to the cost. The two items to be added are a gear and a 10-turn potentiometer. The gear is available for about $7 and the potentiometer for about $9.[1]

Make a potentiometer mounting plate from scrap 1/8 inch aluminum sheet to the dimensions shown in Figure 1. Take care to locate the two holes accurately. A good source of 1/8 inch aluminum for your projects is salvaged front panels from commercial gear. These are available at hamfests for practically nothing. Cut the plate to size with a hacksaw or jigsaw and drill the mounting holes. File the edges smooth and square to dress up your work and sand the plate to deburr it and remove any old paint. See the sidebar for metal-working tips.

Using the control box, rotate the rotator CW to the stop. Remove all the external mast mounting hardware from the rotator and save it for future use, including the 1/4-20 mast-mounting studs. Gently remove the weather seal. The rotator is held together by three black finished hex-head bolts. One of the bolts is under a fiber insulator. Do not remove the insulator. Just bend it up slightly to allow access to the third bolt. The bolts are torqued down heavily. Use a 3/8 inch socket to remove them. Remove the gear mechanism from the housing.

Rotate the 10-turn potentiometer CW to the stop and mount the potentiometer to the plate. Place the gear on the potentiometer

[1]Notes appear on page 6-12.

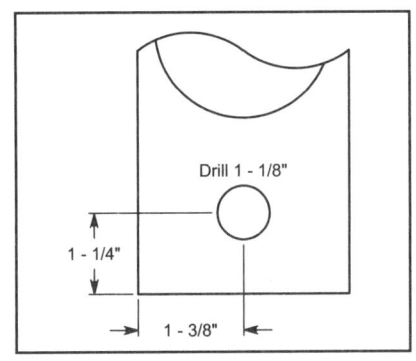

Figure 3—Cut a hole in the rotator housing to clear the potentiometer. This hole is difficult to measure. Drill a small hole after the location is marked to check the placement. Then punch a 1 1/8 inch hole and check for clearance. Finally, epoxy a 1 inch PVC pipe cap over the potentiometer and paint.

shaft, with the hub away from the potentiometer. Then mount the plate through the existing hole, as shown in Figure 2, using a 1 1/2 inch 1/4-20 machine bolt, with extra nuts as spacers. Align the gear to engage with the existing drive gear. Once the gear is properly engaged, torque the mounting nut down until the split lock washer is fully compressed to prevent any movement of the plate. Check the gear again.

Wire the potentiometer, and route the wires down to the terminal area so that moving parts do not trap them. A dab of clear RTV will hold them in place.

Connect the rotator to the control box and rotate the rotator counterclockwise. Observe that the potentiometer gear is rotating freely. Run the rotator into the counterclockwise stop. It will be necessary to support the rotator main shaft when the rotator is turned outside the housing. The potentiometer gear should slip on the pot shaft when the pot stop is reached. The potentiometer should indicate zero just as the clockwise actuator stop is reached. Rotate the actuator from stop to stop and ensure the 10-turn potentiometer is not hitting its stops. When the potentiometer is positioned correctly, put one drop of superglue on the pot shaft and gear hub. Be careful not to get it on the potentiometer bushing, which will lock the shaft.

Cut a hole in the rotator housing to clear the potentiometer, as shown in Figure 3. Note that the walls of the casting slope slightly. Measure the location of the hole from the outside of the casting with it resting on a flat surface. Use a square to project the corner into the plane of the top of the housing and then measure from the blade of the square. Otherwise the slope will throw the hole location off a little. A couple of triangles will help. If the hole is just slightly off, it is big enough to clear the potentiometer. Mark the hole and drill it with a small drill. Then make a trial assembly of the gear mechanism into the housing and check to see if your small hole is in the center of the potentiometer. If not, move it. When you get it right, drill or punch out the small marker hole to 1 1/8 inch. If you do not have a way to drill or punch the hole, drill the biggest one you can and enlarge the hole to 1 1/8 inch with a round file. Check, from time to time, to see that the potentiometer will be centered in the hole when it is mounted. Notice that the solder lugs will also have to clear.

Assemble the gear mechanism into the housing. Check the pot wiring to ensure it is clear of rotating parts and not crimped under something. Reinstall the black bolts and torque them down securely. Make certain the shaft is not binding on the housing and adjust the position slightly before tightening the black bolts.

Cement a 1 inch PVC pipe cap over the potentiometer, using clear epoxy cement.

Antenna System Projects   6-7

Check the rotator operation and the potentiometer wiper resistance. It should vary from nearly zero to about 485 Ω as the rotator goes from stop to stop. If the potentiometer resistance stops changing before the rotator hits the stop, something has gone awry and should be corrected before proceeding.

When everything has been checked, paint the rotator and the PVC pipe cap. Both the PVC and the epoxy cement will deteriorate in the sun unless they are painted. Besides, a painted rotator will make the job look more professional.

Building the Control Box

The control box has two simple jobs to do:
- Power the rotator motor in the correct direction.
- Convert the potentiometer resistance to a display in degrees.

Although a few rotators were manufactured with dc motors, all the units currently in production use split-phase ac motors. A reversible split-phase motor has two identical windings. One winding is fed with 30 V ac. The other winding is fed 30 V ac shifted in phase by 90°. The result is a rotating field with both starting and running torque. To reverse direction, simply change the windings that have the respective voltages. The 90° phase shift is obtained by putting a large value capacitor in series with the winding. A DPDT center-off switch will control the motor direction.

The voltage on the wiper of the pot represents the position of the rotator. If the voltage across the pot is adjusted so that a full rotation of the rotator represents 3.60 V, a voltmeter will read directly in degrees. A small error at zero will result because the resistance of the ground wire from the rotator to the control box is in series with the pot. The error resulting from the simplified version of the circuit is only a degree or so, at worse, with a reasonable length of rotator cable. If the zero error is unacceptable, several ways exist to null it out. These will be discussed later.

The digital panel meter used in the prototype is a C+C Model PM-1029B.[2] The C+C series of meters are based on the Intersil 7107 IC chip. Built in China, the meters are simply copies of the Intersil application note circuit. They are low cost and accurate and have numerous features including range selection, automatic polarity display and independent selection of the decimal point. The unit chosen requires +5 V dc at approximately 80 mA. The LED characters are 0.84 inch high, resulting in

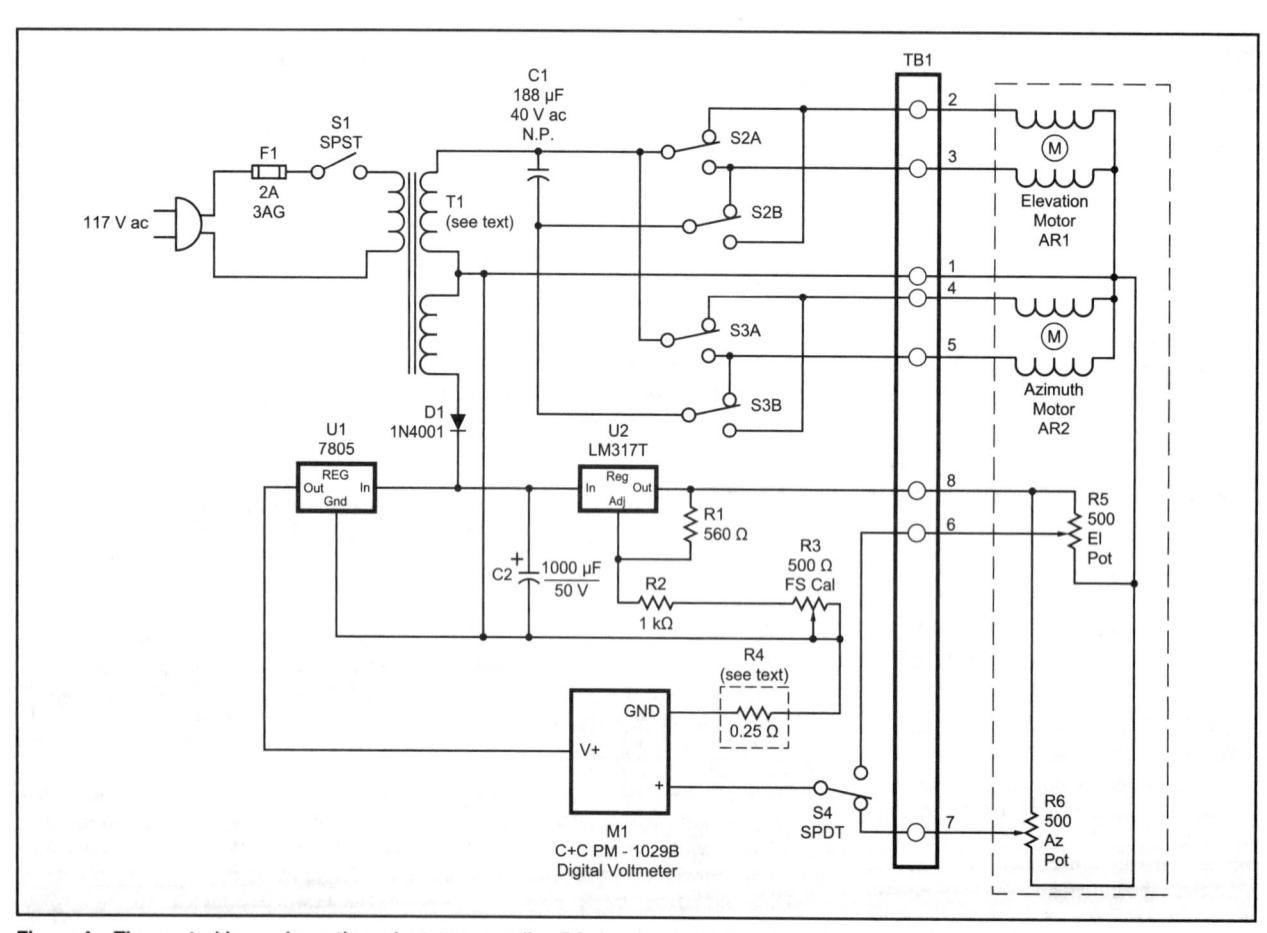

Figure 4—The control box schematic and rotator parts list. RS denotes RadioShack part numbers. Mouser Electronics, 1000 N Main St, Mansfield, TX 76063; tel 800 346-6873; www.mouser.com/.

AR1, AR2—Antenna rotators, RS 15-1245.
C1—188 μF, 40 V ac, non-polarized (salvaged from old rotator control box).
C2—1000 μF, 50 V, RS 272-1047.
D1—1N4001, 1 A, 50 V, RS 276-1101.
F1—Fuse holder, RS 270-364.
M1—Voltmeter, digital, C+C, PM-1029B (see text, Note 2).
R1—560 Ω, 1/4 W, RS 271-1116.
R2—1000 Ω, 1/4 W, RS 271-1321.
R3—500 Ω potentiometer, Mouser 31VA205.

R4—0.25 Ω (see text) #30 AWG wire on 1/2 W resistor.
R5, R6—500 Ω, 10 turn potentiometer, Mouser 594-53611501.
S1—Switch, SPST, toggle, RS 275-651.
S2, S3—Switch, DPDT, toggle, RS 275-709.
S4—Switch, SPDT, toggle, RS 275-652.
T1—Transformer, 117 V ac pri, 2@24 V ac, 1 A sec (salvaged from old rotator control box).

TB1, TB2—Terminal block, 8 screws, RS 274-670.
U1—Voltage regulator, 7805, RS 276-1770.
U2—Voltage regulator, LM317T, RS 276-1778.

Misc
Gears—2 @ 60 teeth, 48 pitch Delrin gears, Small Parts, Inc GD-4860 (see Note 1).
Enclosure—Project box, RS 270-274.

an easy to read display. The meter sells for about $14. The scale of the panel meter is set at 199.9 mV at the factory. The control box requires a meter with a full scale of 19.99 V. Shorting pairs of pads on the circuit board sets the scale. Carefully remove the solder on the 200 mV pads, and put a small blob of solder on the 20 V pads. Connect the +5 V power supply to the V+ pad. Connect the GND terminal to ground, and the input voltage to be measured to the V_{IN} terminal. The V– terminal and decimal point selection pads are not used.

The schematic of the control box is shown in Figure 4.

One winding of the transformer supplies the power for the motor, and the other winding is used to build a small dc power supply to excite the potentiometer circuit and supply 5 V to the meter.

The control box is built into a Radio Shack enclosure measuring approximately 3 inches tall × 8 inches wide × 6 inches deep. The enclosure is thin gauge steel, difficult to work, but producing a good-looking result. The front panel was doubled with a 1/16 inch thick piece of aluminum to make the box sturdier when the switches are flipped. The doubler is probably not necessary. It does have the advantage of providing a pattern for the front panel, which could save a misdrill in the expensive box.

The transformer, line cord and motor capacitor were all taken from the original control box. The switches and miscellaneous parts were all purchased new. The switches have large paddles and are easy to flip. However, the paddles are plastic and may not last as long as regular metal bat-handled toggle switches.

The first thing to do before any construction is started is to put two layers of masking tape on the front and rear panels to preserve the paint. Then, drill the panels as shown in Figure 5. Deburr the holes carefully. The cutout for the meter was done with a jigsaw, then dressed with a file. Modify the cutout to fit your meter.

When all the holes are cut and deburred, apply the symbolization label shown in Figure 6. The label is made from clear sheet Mylar label material obtained at an office supply store. Print the label sheet on an ink jet or laser printer. Be careful to get the label positioned perfectly with respect to the holes and free of wrinkles or bubbles. Use a hobby knife to cut away the Mylar in the holes. Mount the front panel and rear panel components first, because they will not fit into place after the transformer is mounted. Terminal strips mounted on the transformer hold most of the miscellaneous electrical parts. See Figure 7. The voltage regulators are mounted on a small angle bracket, also on the transformer.

When everything is mounted, wire the box with #22 AWG or larger stranded wire. Train the wires into cable runs to keep the interior as neat as possible. Use cable ties to secure the wiring. The prototype was wired with military surplus high temperature wire in several colors.

Mounting and Wiring the Rotators

The elevation rotator can be mounted directly to the azimuth rotator by using a small adapter plate made of 1/8 inch aluminum as shown in Figure 8. Build one piece, sand to debur it and paint.

Remove the 1/4 inch mast-mounting studs from the elevation rotator body with a pair of vise grips. Bolt the adapter plate to the rotator with 1/4-20 machine bolts 1/2 inch long. Use split-lock washers under the bolt heads and tighten until the washers

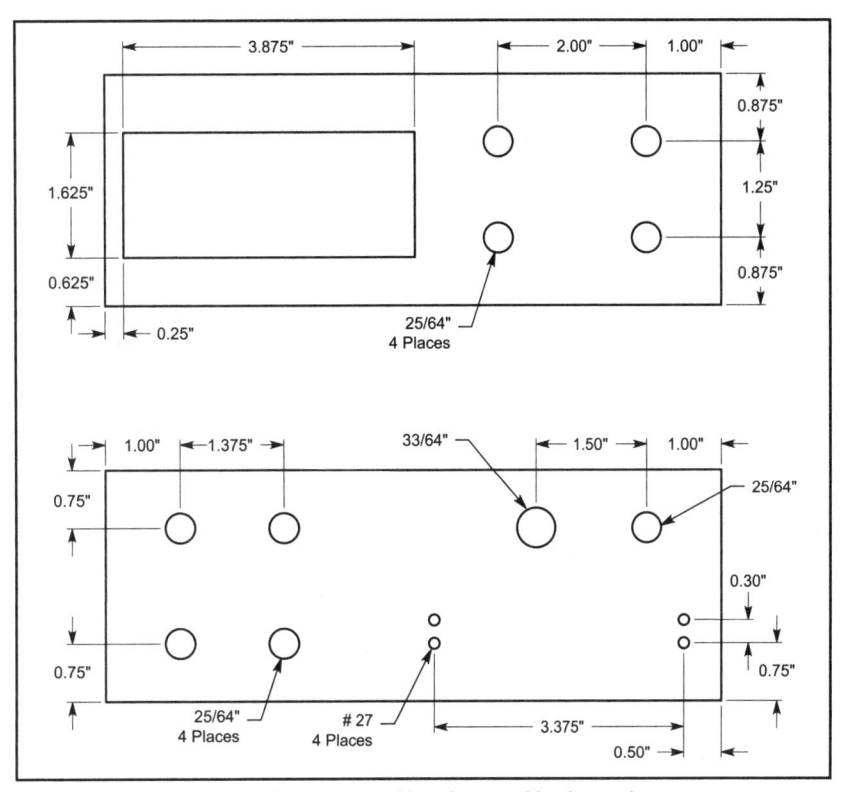

Figure 5—Drilling template for the control box front and back panel.

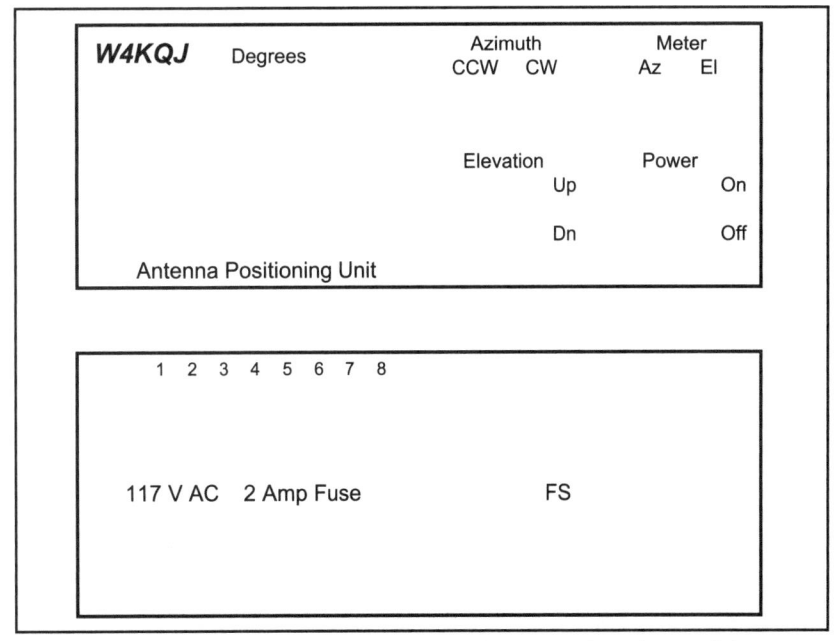

Figure 6—Control box labeling. The label can be printed on a clear sheet and then applied to the box.

Antenna System Projects 6-9

are compressed.

Cut the U-bolts on the azimuth rotator shorter by ½ inch. Set the mast brackets aside. Mount the elevation rotator on the azimuth rotator shaft by the shortened U-bolts. Start all four nuts, and then tighten the bottom nuts until the split lock washers are compressed fully. Then tighten the top nuts. There will be a slight gap from the plate to the rotator shaft because the shaft is tapered slightly. Do not tighten the nuts so tightly that the plate is deformed.

A 30 inch piece of flat, 5-wire cable was used from the elevation rotator down to the azimuth rotator. Strip ¼ inch of insulation from one end of each of the wires. One wire will be silver colored and the rest will be copper colored. Put a red crimp-on lug on the silver wire and the next two wires. Put a red wire splice on the last two wires. Connect the silver colored wire to terminal 3 in the elevation rotator and the next two wires in order to terminals 2 and 1, respectively. Splice the outside wire farthermost from the silver wire to the wire going to the top of the pot. Splice the remaining wire to the wire going to the wiper. Connect the wire from the bottom terminal of the pot to terminal 3 on the rotator. Bring the 5 wire cable from the elevation rotator through the grommet and then into the azimuth rotator to an 8 screw terminal block mounted to the lid. Put red crimp-on terminals on all five wires. Connect the cable lugs to the correct screws on the terminal block as shown in Figure 9. Connect the wires from the azimuth rotator to the remaining screws. Remember that the elevation rotator will have to rotate through 360°, and leave a slack rotator loop in the cable.

The rotator assembly is connected to the control box by standard 8 wire rotator cable. The rotator cable will have two wires larger than the others. On the CDE rotators these wires were used for the ground and the brake solenoid. On this project, one is used for ground and one for the potentiometer voltage. Connect one of the two large wires, regardless of color, to pin 1 on the block. Connect the other large wire to pin 8. Connect the remainder of the wires in color code order to pins 2 through 7. Run the cable through the slot and replace the lid. The grommet is much too small for both cables, so leave it out and seal the wire entry with a glob of RTV sealant.

Connect the 8 wire cable from the az-el assembly to the control box. Be sure to use the same color code in wiring the box. Test the whole project before mounting the antennas up on the tower or mast.

Rotate the azimuth rotator clockwise to the stop. Do not continue to apply power after the stop is reached. Set the FS potentiometer (R3) for a 360° reading on the display. Rotate the azimuth rotator counterclockwise to the stop. Check the zero reading. If the zero reading is too high, see "That Pesky Zero Error," following. Now rotate the azimuth rotator to 180° and note that the arrows on the rotator housing and rotator shaft line up. These arrows will be used to set the azimuth axis to SOUTH with a compass after the assembly is in place on the tower or the mast. In a similar manner, check the calibration of the elevation axis. There is no independent adjustment for the elevation axis.

Antenna Mounting Options

If the antenna array is small and light, the elevation mast can be mounted directly to the elevation shaft, as shown in Figure 10. The antennas will then be mounted all on one side. If the wind loads reverse-drive the rotator, due to the lack of a brake, the position pot will still continue to indicate correctly and the angle can be reset.

If a larger antenna array is used, the antennas can be mounted on either side of

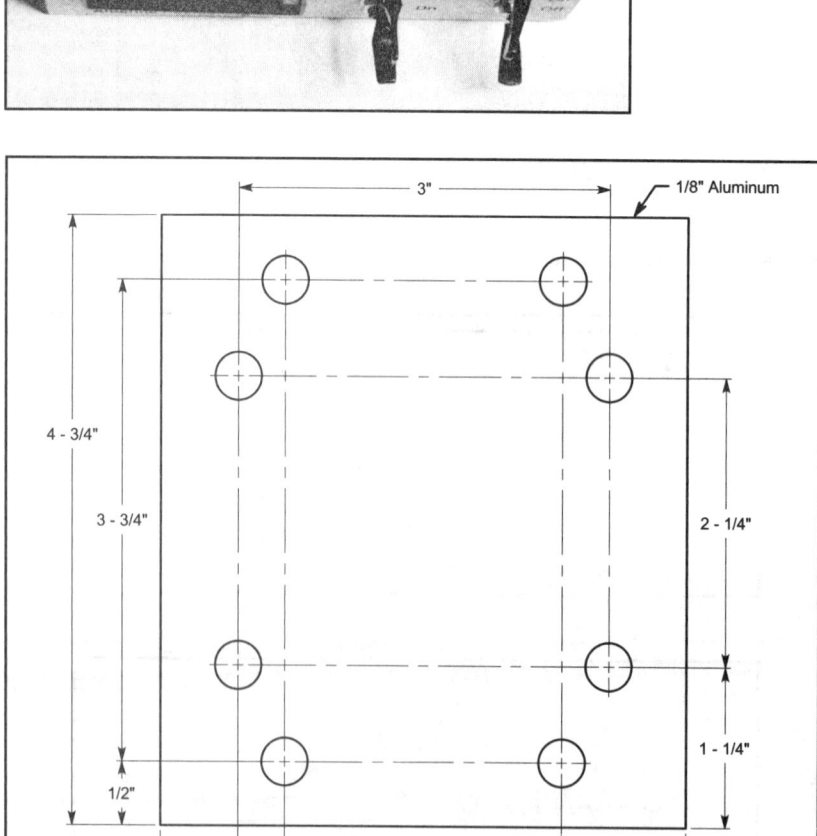

Figure 7—Control box component mounting. The components are mounted on terminal strips secured to the transformer mounting screws. Two screws mount the transformer on diagonal corners; the other corner screws secure the terminal strips.

Figure 8— Adapter plate to secure the azimuth rotator to the elevation rotator.

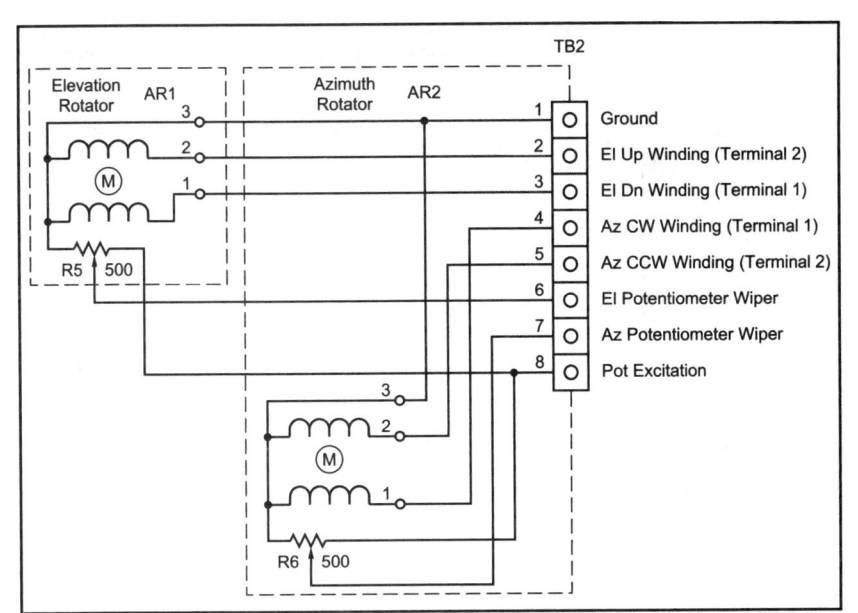

Figure 9—Rotator terminal block wiring. The elevation rotator is wired with 5-wire rotator cable, terminated at a terminal block in the azimuth rotator. The combined unit connects to the control box with 8-wire rotator cable. See Figure 4 parts list.

Figure 10—A single antenna mounted to one side. The rotator is ideal for positioning an S band antenna for receiving the AO-40 satellite.

Working Sheet Metal

Aluminum sheet metal is easy to work, but a few tips will aid you in turning out professional looking work. Mark the outline of the work piece with a square and a sharp scribe. If you do not have a metal shear, cut sheet metal with a saw where possible, as hand shears will bend the edges. The best setup is a radial arm saw with a metal cutting blade installed. Lacking that, a hacksaw with a fine-toothed blade can be used.

A couple of pieces of ¼ × 1 inch aluminum stock about a foot long are great tools. Sandwich the sheet metal work piece in a vise between the two tool pieces with the cut line barely exposed. The tools will make the sheet metal rigid and easy to cut. Use a fine file while the sandwich is still clamped to smooth out the saw marks. Sand the edges smooth and straight with a belt sander or a piece of sandpaper laid flat on the workbench. Be very careful to hold the work at 90°. A sloping edge is a dead giveaway of novice work.

When the work piece is sanded to the correct outline, mark the position of the holes with a square and sharp scribe. Lightly center punch the hole locations. Drill a small pilot hole. Use of a small first hole will insure better accuracy. If at all possible, use a small drill press and clamp the work to the table, using a scrap of material between the work piece and the clamp jaw. Clamp brass anytime you work with it. Even a small drill will turn brass sheet into a lethal weapon.

To deburr and finish the work, sand the flat sides of the aluminum sheet with a small power sander. Use ordinary 120 grit wood sandpaper. Finish with 400 grit wet metal sandpaper for a shiny surface. Use a small sanding block if you do not have a power sander. Do not use a larger drill to deburr holes. The marks will show. Sand the flat surface until the burr disappears.

the rotator as shown in Figure 11. Build the mast assembly shown in Figure 12. Use two 1 inch galvanized steel Ts connected with a 3 inch threaded nipple. The mast is then slid through one T and locked in place by the ¼ inch machine bolt tapped into the T. Do not use 1 inch threaded galvanized pipe for the mast. It is too heavy. Obviously, the elevation mast will hit the structure at two rotation angles, but the elevation angle covered is more than adequate. Try to keep the assembly light and as well balanced as possible. Paint the entire assembly. The galvanized pipe will rust where it was threaded.

When the antennas are all mounted, align the azimuth rotator to SOUTH with the indicator reading 180°. Check the setting of the zero and 360° positions and correct as required using the FS potentiometer. Set the elevation angle with a level so that the indicator reads correctly at 0° and at 90°.

That Pesky Zero Error

It was mentioned earlier that a small error at zero exists. This is because the resistance of the ground wire from the rotator to the control box is in series with the end resistance of the pot. The resulting error is only a degree or so, at worst, with a reasonable length of rotator cable and with the pot properly set so that the end resistance is small. However, if the error is deemed unacceptable, two methods of correction are possible. The easiest is to insert a very small resistance in series with the ground of the meter, as shown by the dashed lines in Figure 4. The resistance will be on the order of 0.25 Ω. Parallel wire 1 Ω resistors until the meter reads zero at zero rotation or use a small resistor made from #30 AWG magnet wire wound on a large value ½ W resistor body. The resistor value must be kept very low or the excellent zero stability of the voltmeter will be compromised. The resistor represents positive feedback and if too large in value, the meter will oscillate and never settle down.

A much better but more costly method is to establish a bridge circuit with the position potentiometers on one side and a zero pot on the other. The bridge is balanced at zero rotation and the meter then reads exactly zero. To implement this zero circuit, the plated through hole on pin 30 of the IC must be drilled out so that a wire can be attached to it. The plated through hole is accessible, but extreme care must be used due to the board layout. The schematic of the zero correction circuit is shown in Figure 13. I recommend living with the small error and not implementing either fix.

Growth Potential

These days everyone wants computer-controlled equipment. For AO-40, computer control is not necessary. The satellite apparent motion is so slow that an occasional tweak of the antenna is all that is required.

Antenna System Projects 6-11

Figure 11—A balanced antenna mast. Heavier arrays should be balanced on both sides of the rotator to equalize the loading. As the elevation mast can't be run through the rotator, a pair of pipe "T"s are used.

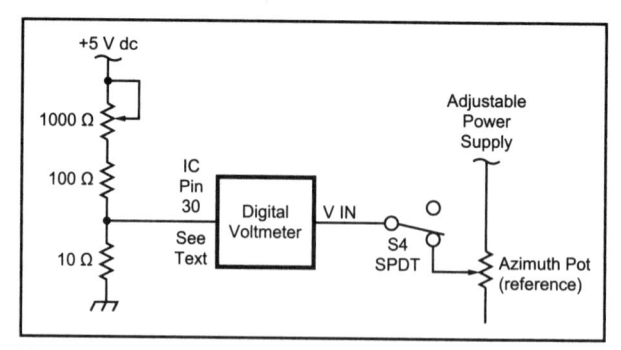

Figure 13—Control box zero error adjustment. If the small zero error is unacceptable a modification is possible. See the text.

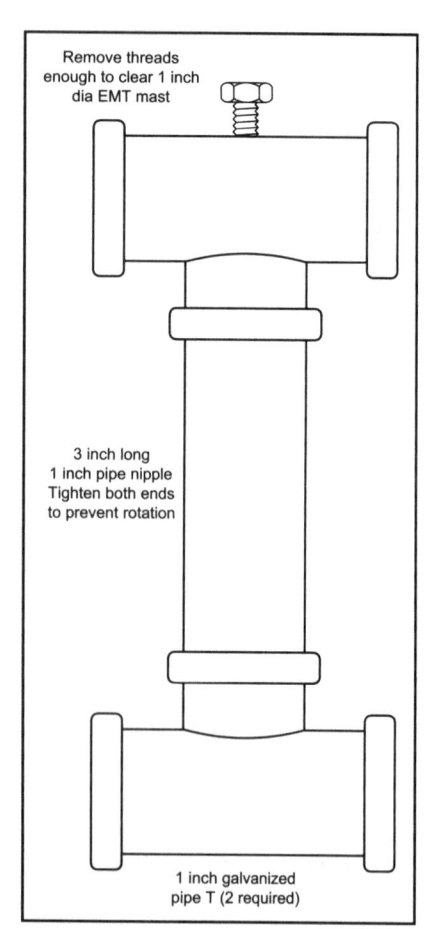

Figure 12—The pipe "T" detail. Two "T"s and a 3 inch nipple allow the mast to be balanced to equalize loading. The top "T" has a machine bolt to lock the mast.

For polar satellites, however, computer control would be a nice feature. To add computer control to this system, one could use any of the popular interfaces already described by others. However, adding even the simplest interface increases the project cost. The simplest, lowest cost interface is the *Fodtrack* designed by Manfred Krohmer, XQ2FOD, and distributed by *AMSAT-CE*. The printed circuit board is available from FAR Circuits.[3]

Several modern tracking programs have a *Fodtrack* driver option, including the software sold by *AMSAT-NA*.[4]

The *Fodtrack* board has four open collector outputs representing each of the directions of rotation of the rotators. Adding four small SPST relays will enable the circuit to drive the rotators. The contacts are simply paralleled across the respective switch contacts. The output of the two position potentiometers will drive the *Fodtrack* board without additional circuitry.

If that interface is added, mount the printed circuit board behind the voltmeter on spacers to the bottom of the box.

The beauty of the interface is that the meter display will continue to function. The angle displayed by the computer can be compared to the angle displayed by the meter, in order to assure that the antenna is really pointed where you think it is.

Notes
[1]The gears used in this project are available from Small Parts Inc, 13980 NW 58th Ct, Miami Lakes, FL 33014-0650; **www.smallparts.com**. Part no. GD-4860. $6.85 each plus shipping and handling.
[2]The C+C voltmeters are available from Circuit Specialists, 220 S Country Club Dr #2, Mesa, AZ 85210; **www.web-tronics.com**. Part no. PM 1029B. $15.95 plus shipping and handling.
[3]The *Fodtrack* printed circuit board is available from FAR Circuits, 18N640 Field Ct, Dundee, IL 60118-9269; **www.farcircuits.net**. Price $4.50 plus shipping and handling.
[4]*AMSAT-NA* and *AMSAT-CE* distribute *Fodtrack* software on the Web as freeware. The distribution package contains instructions for building the interface board. For more information see the extensive discussion of *Fodtrack* interfacing by Jesse Morris, W4MVB, in *The AMSAT Journal*, Mar/Apr 2002.

Lilburn R. Smith, W5KQJ, was first licensed in 1956. He holds an Extra Class license and has a BSEE degree from Texas Tech University. Lilburn has been involved in microwave, VHF and laser design and holds one US patent. He is a past president of the Central States VHF Society and the North Texas Microwave Society, and can be reached at 290 Robinson Rd, Weatherford, TX 76088; or via e-mail at **W5KQJ@arrl.net**.

Circularly Polarized Yagi Antennas for Satellite Communications

Jim shows us how he makes satcom antennas on a semi shoestring budget.

Jim Kocsis, WA9PYH

Back in 1985 my big interest was VHF weather satellites. To get a really good, clear picture as the satellite neared the horizon you needed a circularly polarized (CP) Yagi antenna designed for a frequency of 137 MHz and pointed at the satellite. I built one and got rock crushing signals until the satellite went over the horizon — it worked great!

Amateur Satellites Appear on the Horizon

Around 2000 my interest turned to VHF and UHF amateur satellites. After several years of mediocre performance from a single seven-element 2 meter Yagi, I decided to improve my antenna situation. I decided that I needed CP Yagis for 70 cm and 2 meters to get good signals from the SSB low earth orbit (LEO) satellites.

In this article I will show you how I built my own CP Yagis for much less than the cost of new antennas.

Why Circular Polarization?

CP is very effective at eliminating signal fading both to and from the satellites. There are two ways that CP eliminates signal fades:

- It mitigates fading due to spin modulation that occurs as the satellite turns.
- CP eliminates the effects of Faraday rotation as signals pass through the ionosphere.

If a linearly polarized antenna is used on the satellite and the ground station, the signal will fade as the satellite polarization rotates (vertical through to horizontal). I witnessed this firsthand years ago while trying to hear AO-10 with a single plane 2 meter Yagi. When the polarization of the signal matched my antenna, signals were strong. When the signal polarization rotated, the signal dropped out completely — only to return again in a few seconds. With a CP antenna there is much less fading because any rotation results in a constant signal level.

For a full explanation of each concept see *The Satellite Handbook*, page 4-2.[1]

System Design

See Figure 1 for a view of the basic sys-

[1]Notes appear on page 6-16.

tem configuration of my CP Yagis. Note that I have elected to use a folded dipole as the driven element (DE). This has a feed point impedance of 200 Ω. A 4:1 balun (balanced to unbalanced transition) is used to transform the balanced 200 Ω impedance to 50 Ω unbalanced for a coax feed. This is done using a ½ wavelength section of 50 Ω coax connected as shown. Circular polarization is accomplished by adding a second Yagi to the boom at a right angle to the first and positioned back ¼ wavelength behind the first Yagi.

The two antennas are then connected through multiple ¼ wavelengths of 75 Ω coax to a T connector. The 75 Ω sections perform another impedance transformation. The lengths of the 75 Ω sections and the ¼ wavelength spacing provide the proper phasing of the signals to generate a CP signal. The two 100 Ω feed points are connected in parallel at the T connector to result in a 50 Ω system impedance. It looks like a complex way to get to 50 Ω but it's simple and it works.

I used chapter 9 of *The ARRL UHF/Microwave Experimenter's Manual*, pages 9-3 through 9-8 to determine the element

Antenna System Projects   6-13

lengths and spacing along with the dimensions of the DE for a 70 cm CP Yagi, with elements electrically connected to and side mounted on the boom.[2] If you use the same element diameter, boom size and mounting methods for each antenna you can use the dimensions shown in Tables 1 and 2.

Materials Needed

There are no impossible to find parts in these designs. If you have a well-stocked junk box of mechanical parts, attend hamfests and have one or more local home improvement stores nearby, you can also build these antennas at low cost.

First things first. These antennas will be much easier to build if square tubing is used for the boom. It is very difficult to drill holes at an exact location on a curved surface unless you have the needed mechanical skills and equipment.

I used insulated through the boom construction for the 2 meter Yagi (see Figure 3) and on the boom construction on the 70 cm antenna (see Figure 4). The dimensions in Tables 1 and 2 reflect those methods using a ¾ inch square boom. For the through the boom method, shoulder washers are needed to insulate the elements from the boom. Nylon shoulder washers are available at a very low price from MSC (**www.mscdirect.com**). Insulating washers made from Celcon are preferred as they withstand UV (sunlight) better, if you can find a source. Tubing in ¾ inch size is available from **www.onlinemetals.com**. They also have ¼ and ⅜ inch diameter round tubing available.

The DEs for 2 meters are built using ⅜ inch diameter aluminum tubing. The DEs and all directors and reflectors for 70 cm, are built of ¼ inch diameter aluminum tubing. One source I found is MSCDirect. I also found ¼ inch tubing for my antennas available locally in 12 foot lengths. Check the classified directory in your area for suppliers of the aluminum tubing. Lowe's offers ¼ inch solid aluminum rod in 36 inch lengths for $3. This will work for a 70 cm antenna but will make it heavier. The solid rod elements will, however, resist bending much more than tubing.

The coaxial connectors, 75 Ω and hard line coax are available at many hamfests I've attended recently. The Type N bulkhead connectors must be able to accept solder. A silver finish is required. I've seen many at hamfests that are very shiny. The surface is not silver and the connectors may not work in this application since solder will not adhere

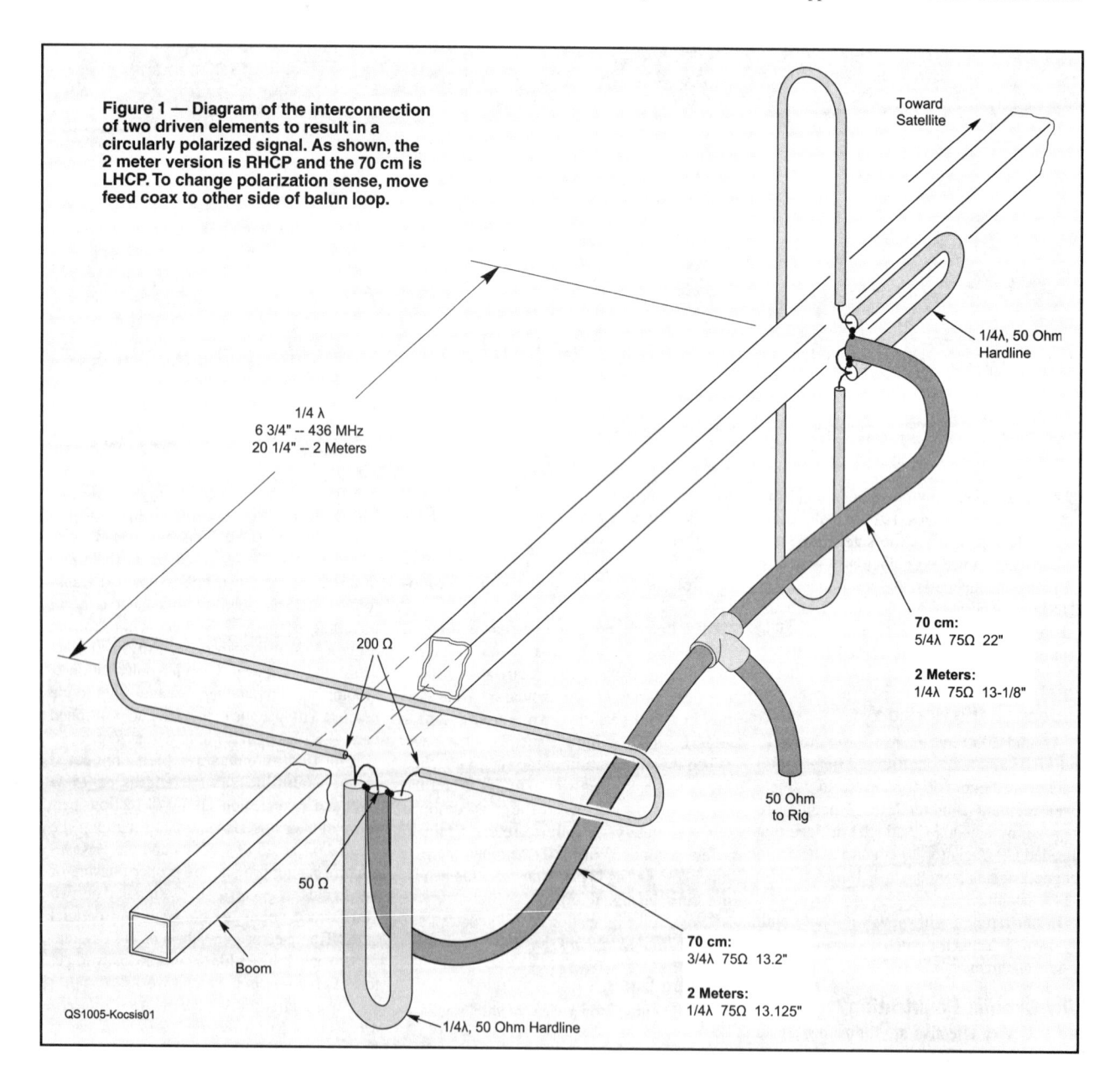

Figure 1 — Diagram of the interconnection of two driven elements to result in a circularly polarized signal. As shown, the 2 meter version is RHCP and the 70 cm is LHCP. To change polarization sense, move feed coax to other side of balun loop.

6-14   Chapter 6

Table 1
2 Meter Seven Element Yagi Dimensions (inches)

Element	Length	Distance from Reflector
Reflector	48 5/8	
Driven Element	38 3/8	15
Director 1	36 1/2	21 3/8
Director 2	36 1/8	35 15/16
Director 3	35 3/4	53 3/8
Director 4	35 5/16	73 3/4
Director 5	35 1/16	96 1/2

Table 2
70 cm 10 Element Yagi Dimensions (inches)

Element	Length	Distance from Reflector
Reflector	16 5/8	
Driven Element	12 3/4	5
Director 1	12	7 3/16
Director 2	11 7/8	12
Director 3	11 13/16	17 7/8
Director 4	11 11/16	24 11/16
Director 5	11 1/2	32 5/16
Director 6	11 7/16	40 1/2
Director 7	11 1/4	49 1/8
Director 8	11 3/16	58 1/16

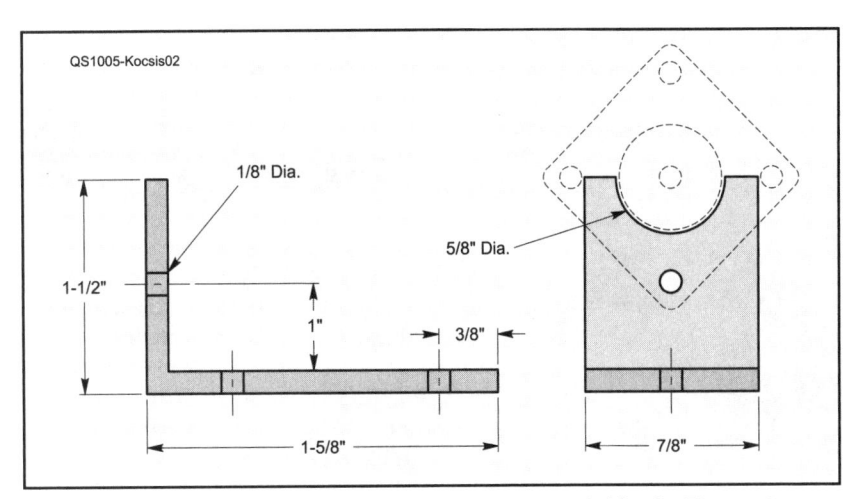

Figure 2 — Fabrication details of the mounting brackets needed for the 70 cm antenna. Two are required.

to them. Used silver connectors may be tarnished but they will work fine as long as there is no damage to the center pin, insulation or threads. I have listed a source on the QST In Depth Web page that provides non silver connectors that will accept solder.[3]

Construction Steps

The assembly of the first plane is easy. Adding the second is a bit more difficult since the first plane of elements will get in your way. Drill all the holes for both planes before you begin assembly. Add the DE (minus the balun and connector) last as it has parts that can be damaged during installation. If at all possible, get a helping hand to hold the antenna while you're adding elements or suspend it from the ceiling inside your garage with rope at one end while providing support at the other end from underneath to position it at a comfortable working height.

With the offset between the antennas and the element spacing it is likely that there will be a conflict between an element in one plane and an element in the other plane with both needing to occupy the same location on the boom. Of course that's not possible, so the solution is to move one ahead a very small distance and the other back a small distance. There will not be a noticeable loss in performance with these small changes in position.

Minimizing Waste and Securing Elements to the Boom

Once you've decided on a tubing source, determine what reflector and director lengths can be cut from each length of tubing to minimize waste. Make a chart of the calculated lengths such as the one in Table 3 and switch things around until waste is minimized. Table 3 shows the lengths I cut from each 12 foot length of ¼ inch tubing to minimize waste. Cut them to length and mark the center of each element clearly with a pencil or marking pen.

Making the Driven Element

The folded dipole DE construction is the most difficult part of making this Yagi. Calculate the total length of tubing required (including the radius at each end) then add 4 to 5 inches. Mark the center point of the DE. Before making the first DE, practice bending some scrap pieces of tubing so that you will

Figure 3 — Elements mounted to boom of 2 meter antenna. The director with Celcon insulator and stud/rod detent washer are clearly visible, as is the driven element feed point with plastic plate.

Figure 4 — Detail of a 70 cm director mounted to the boom showing clamps, washers and metal tapping screws.

Antenna System Projects 6-15

Table 3

Minimizing Waste from 10 Foot Sections for 2 Meter Elements (inches)

Section 1		Section 2		Section 3		Section 4 (90")	
35¾	Director 3	36½	Director 1	48⅝	Reflector	48⅝	Reflector
35¾	Director 3	36⅛	Director 2	35⁵⁄₁₆	Director 4	36½	Director 1
35⁵⁄₁₆	Director 4	36⅛	Director 2	35¹⁄₁₆	Director 5		
13 waste		11¼ waste		1 waste		5 waste	

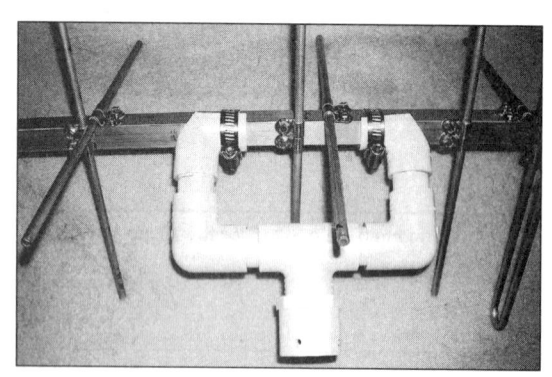

Figure 5 — View of the PVC support assembly for the 70 cm antenna.

know where to start the bend so the completed bend is located properly. Note how it slips through the bender and slightly changes the point at which the bend occurs compared to where you expected it.

This is a very important point. This slippage will affect the final length of the DE after making the second bend. Allow a few extra inches of tubing at the end of the first bend then put the tubing in the bender for the second bend. Another very important point — make sure that the second bend is in the same plane as the first bend! If the second bend is out of the plane, the DE will be twisted and become scrap – not that I ever did that.

The 4:1 Balun

The balun is constructed from RG-141, 50 Ω hardline, for 70 cm and RG-59/U for 2 meters. All junctions are coated with a sealer to keep out water and dust.

Notice that on the 70 cm antenna the feed line going to the coax T fitting exits the antenna at the rear and on the 2 meter antenna the feed line exits forward toward the support cross boom. I did this because the coax and T junctions are large compared to a wavelength at 70 cm but smaller (about ⅓ the size) at 2 meters. I routed the feed lines this way because I felt that the presence of a large object in the plane of the antenna could possibly distort the pattern at 70 cm but would be less of a problem at 2 meters. Fabrication details for the brackets needed to support the 70 cm balun are shown in Figure 2.

Phasing Harness and Offset Spacing

This is probably the hardest part to explain, visualize and understand and so deserves some of your time to study and understand it. The purpose of the phasing harness and offset spacing is to feed the signal from the transmitter equally to both halves of the antenna, but phased so that the signal is launched in a circular pattern with the correct sense of rotation. I accomplished it by offsetting the two antennas ¼ wavelength for LHCP then adding a ½ wavelength in one leg to reverse it to RHCP. Figure 1 shows the placement and orientation of all parts that affect polarization sense.

Supporting the CP Yagi Antennas

I've read a few discussions about what happens to the performance of Yagi antennas when a metal cross-boom is used for support. The cross-boom (the long cross pipe that goes between the two antennas) can skew the pattern and will affect the circularity of the waveform of a CP antenna. One solution is to use fiberglass rod or tubing since it is not metal. A single 8 foot length costs over $80 with shipping. A little more research on the Web led me to a site with an article written by Kent Britain, WA5VJB.[4]

Kent ran some experiments to determine the effects of placing a metal mast at various locations along the boom of a single plane Yagi. He also varied the angle between the plane of the antenna and the cross-boom. It is very interesting reading but the bottom line is that a properly placed metal pipe that reaches only the antenna boom (it doesn't go entirely "through" the antenna) and is positioned at 45° to the plane of the antenna will have very little effect on the gain or pattern.

A section of 1¼ inch outside diameter (OD), 1 inch ID, PVC tubing with ⅞ inch OD steel conduit inside for stiffness, is used to support each Yagi near the center of mass to eliminate any torque on the elevation rotator. The steel conduit doesn't extend all the way to the antenna boom — it's spaced a few inches back inside the PVC T fitting so the effect on the antenna should be even slightly less than that shown on Kent's graphs.

I wanted to have a robust structure at the junction of the cross boom and each antenna's boom. Any metal in the field can affect the performance of the antenna so instead of the typical metal plate and four U bolts, I used several PVC 90° fittings, a PVC T fitting and short sections of PVC pipe jointed together with glue. The two 90° fittings at the antenna must be cut off along one side approximately half way so that the boom of the antenna can slip in place. A stainless steel hose clamp at each end grips the antenna boom tightly. Figure 5 shows the 70 cm antenna with this support.

Performance

These antennas provide strong signals from the satellite as it travels all the way down to a few degrees above the horizon. They've been up for more than 4 years, have survived numerous wind gusts above 60 mph, show no sign of failure and work as well as the day they went up.

Notes
[1]S. Ford, WB8IMY, *The ARRL Satellite Handbook*. Available from your ARRL dealer or the ARRL Bookstore, ARRL order no. 9857. Telephone 860-594-0355, or toll-free in the US 888-277-5289; **www.arrl.org/shop**; **pubsales@arrl.org**.
[2]Out of print, but available at **www.amazon.com**.
[3]**www.arrl.org/qst-in-depth**
[4]**www.g6lvb.com/fibermetalboom.htm**

ARRL member Jim Kocsis, WA9PYH, was first licensed in 1964 as WN9LDB. He earned his General class license in 1965 and upgraded to Amateur Extra class in 1986. He is a member of AMSAT and is active on the satellites when not homebrewing something. Jim's other interests are casual DXing, CW on HF, low power CW, especially during Field Day, non-competitive bicycling, cooking and reading travel essays. He has homebrewed small projects his entire ham career, including UHF weather satellite downconverters and antennas.

Jim received a degree in physics from Indiana University in 1976 and is employed as a test engineer at Honeywell Aerospace. You can reach Jim at 53180 Flicker Ln, South Bend, IN 46637 or at **sadiekitty@sbcglobal.net**.

Dual Band Handy Yagi

Thomas M. Hart, AD1B

The popular handheld Handy Yagi can now work on both VHF and UHF.

Recently, I started operating through the SO-50 and AO-51 satellites with my dual band Yaesu FT-60 handheld transceiver. The standard flexible antenna worked, but I decided to investigate handheld Yagi antennas to improve conditions. My goal was to build a simple dual band 2 meter and 70 cm Yagi without driven elements, a matching network or a feed line. In short, my plan was to mount the FT-60 on the antenna as I had with the original 70 cm Handy Yagi.[1]

After testing several configurations on Roy Lewallen's *EZNEC* program (see **www.eznec.com**), I settled on a design with seven directors on 70 cm and three on 2 meters. My Yaesu FT-60 dual band handheld serves as the driven element. Figures 1 and 2 show the configuration.

Construction

The final design balances performance and size. The elements for both bands are interlaced and mounted in parallel. There is no driven element or reflector. Instead, the FT-60 and a bicycle handlebar grip occupy the usual reflector and driven element end of the boom. The 2 meter elements can be rotated parallel to the boom to simplify storage. A screw eye at one end can be used to hang the antenna when not in use.

Computer modeling indicated that $\lambda/4$ element spacing works reasonably well on both bands. This allows the use of a 55 inch boom. The handheld is attached firmly in place by its belt clip. A speaker microphone makes transmitting and receiving very simple. See the illustrations for additional details.

All elements were cut from 1/8 inch diameter steel rod. An article by Ron Hege, K3PF, provided the dimensions for the 2 meter elements shown in Table 1. The 70 cm element dimensions are found in *The ARRL Antenna*

[1]Notes appear on page 6-18.

Figure 1 — Dual band Handy Yagi with 2 meter elements folded for storage.

Figure 2 — An *EZNEC*-generated diagram of the antenna shows the element placement. The Yaesu FT-60 is represented by the half-wave loaded element on the left.

Table 1
Length and Spacing of 2 Meter Elements

All dimensions are in inches.

Director:	D1	D2	D3
Length	37.5	36.375	36.0
Element Spacing	DE to D1	D1 to D2	D2 to D3
Spacing	12	12	12
Cumulative	12	24	36

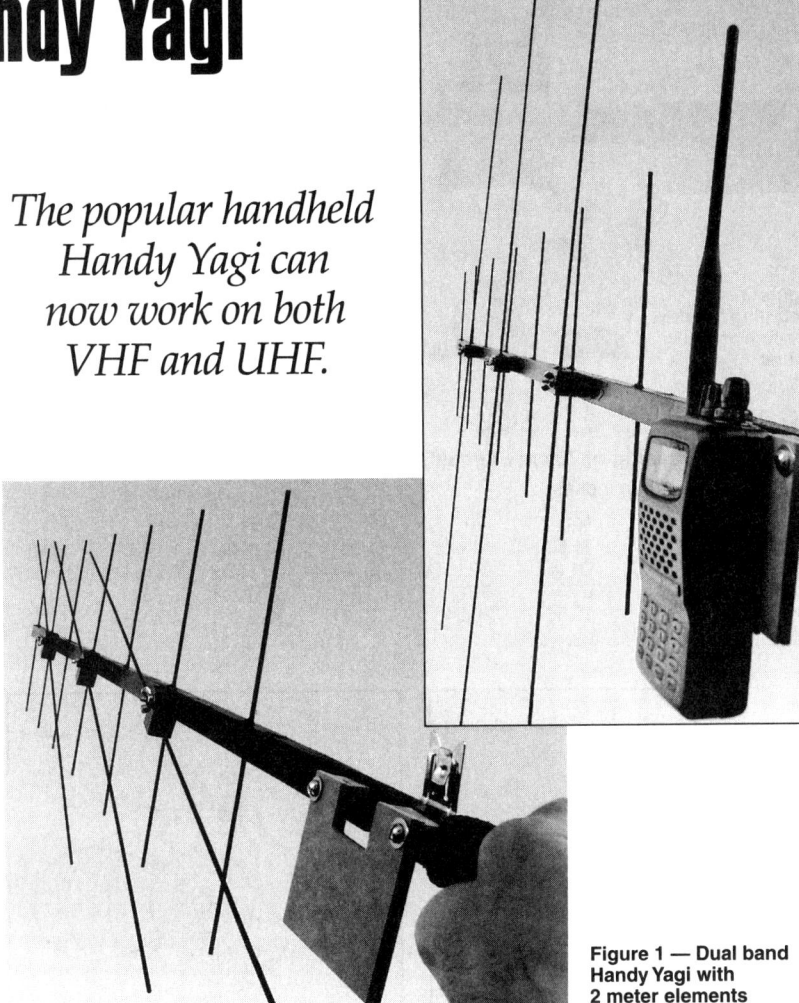

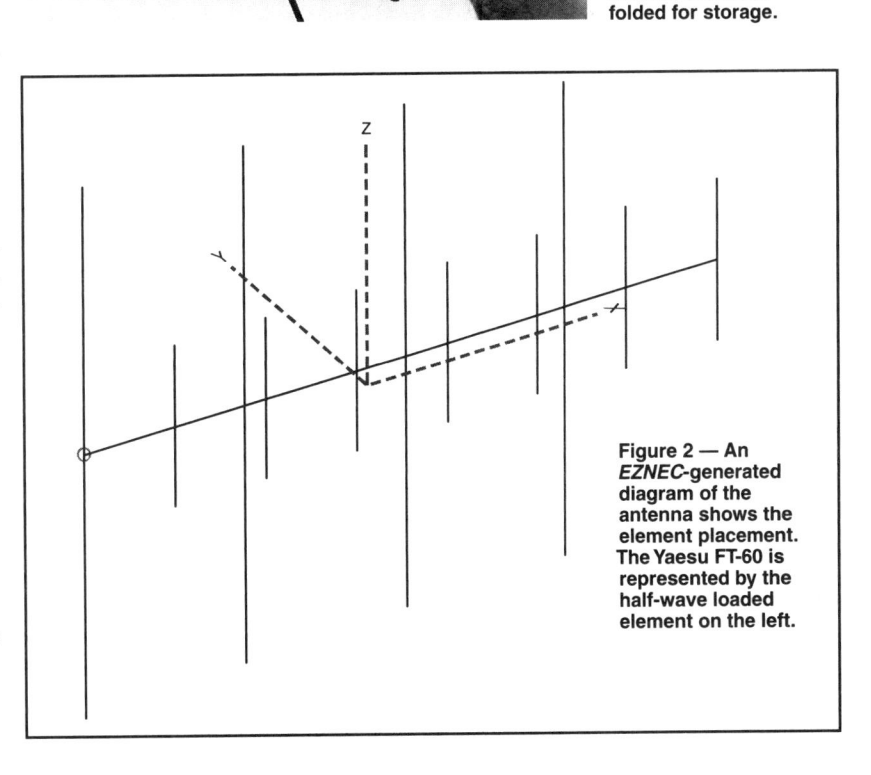

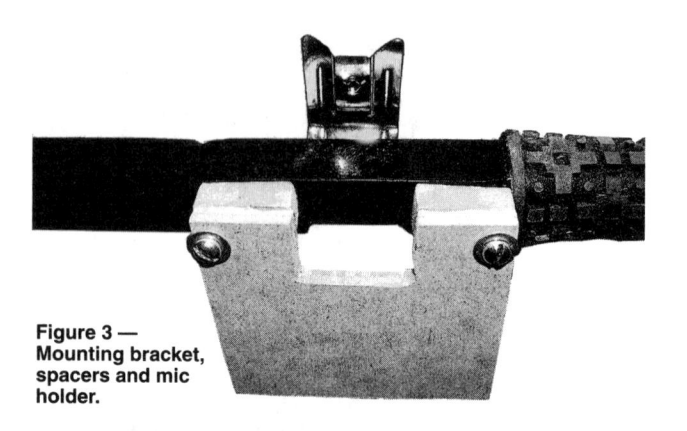

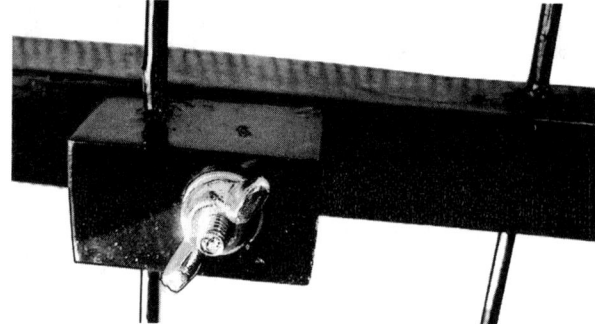

Figure 3 — Mounting bracket, spacers and mic holder.

Figure 4 — The wing nut and bolt allow the 2 meter element to rotate.

Table 2

Length and Spacing of 70 cm Elements
All dimensions are in inches.

Director:	D1	D2	D3	D4	D5	D6	D7
Length	11.750	11.688	11.625	11.563	11.500	11.438	11.375
Elements	DE to D1	D1 to D2	D2 to D3	D3 to D4	D4 to D5	D5 to D6	D6 to D7
Spacing	6.78	6.78	6.78	6.78	6.78	6.78	6.78
Cumulative	6.78	13.56	20.34	27.12	33.9	40.67	47.45

Table 3

Dimensions of Other Antenna Assemblies
All dimensions are in inches.

Boom	0.75 × 0.75 × 55
Handle grip	5
Handheld bracket	3.5 × 3.0 × 0.25 (WHD)
Notch for handheld	1.125 W × 0.75 H
Bracket spacers	0.375
2M element mounts	0.75 × 0.75 × 1.5

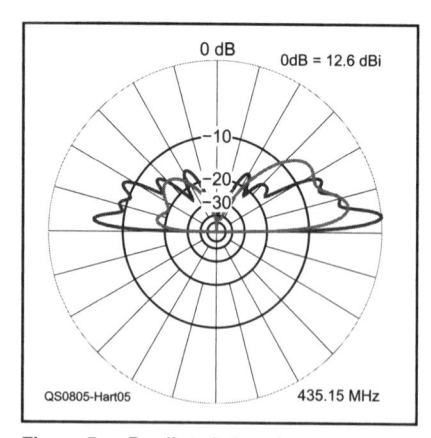

Figure 5 — Predicted elevation patterns of the 2 meter (red) and 70 cm (blue) Yagis.

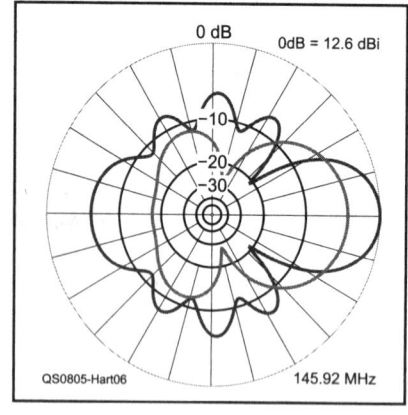

Figure 6 — Predicted azimuth patterns at the peak of the elevation response for the 2 meter (red) and 70 cm (blue) Yagis.

Book and reproduced as Table 2. Other dimensions are shown in Table 3.

Two glued together pieces of ⅛ inch pressed fiberboard form the handheld bracket shown in Figure 3. The elements are held in place with epoxy. After cutting all 10 to length, drill seven holes through the boom and three more through the 2 meter element mounts. Slide the elements into place and apply a bead of epoxy on both sides. Several light applications of epoxy will hold the rods in place.

All 2 meter element mounts require two holes. One secures the element while the second is used for the bolt and wing nut that allow the element to rotate during storage or transport as shown in Figure 4. Finally, two coats of black satin paint give the antenna a more professional appearance.

Testing

Faced with an absence of analytical instruments, the testing process involved *EZNEC* computations and operational observations. *EZNEC* computed the front-to-back ratios as 14 dB for 2 meters and 5.5 dB for 70 cm. *EZNEC* azimuth and elevations patterns are shown in Figures 5 and 6.

Field testing on 2 meters involved contacting the 146.97 MHz repeater in Paxton, Massachusetts. The repeater, according to my *Magellan Topo* 3D GPS software, is 38.0 miles away at a bearing of 251°. The FT-60 helical antenna cannot reach the repeater from my house. With the dual band Handy Yagi, I had no trouble reaching the repeater. Moving the antenna in an arc toward and away from the repeater produced corresponding signal strength changes on the meter corresponding to the predicted directivity.

On 70 centimeters, field tests involved reception of signals from the AO-51 and SO-50 satellites. Signal strength increased dramatically when the antenna approached and centered on the target. My conclusions are that operations on both bands are in agreement with *EZNEC* predictions.

The antenna that I built is designed for satellite contacts. However, the basic concept can be used for portable operations, fox hunting and emergency use as well.

Notes
[1]T. Hart, AD1B, "The Handy Yagi," *QST*, Nov 2007, pp 37-38.
[2]R. Hege, K3PF, "A Five-Element, 2-Meter Yagi for $20," *QST*, Jul 1990, pp 34-36.
[3]R. D. Straw, Editor, *The ARRL Antenna Book*, 21st Edition, p 18-45. Available from your ARRL dealer or the ARRL Bookstore, ARRL order no. 9876. Telephone 860-594-0355, or toll-free in the US 888-277-5289; **www.arrl.org/shop/; pubsales@arrl.org**.

All photos by the author.

Tom Hart, AD1B, began listening to short wave broadcasts in 1961. He received his Novice class license, WN1JGG, in 1968 and has been an active on CW, SSB, RTTY, FM and packet ever since. Tom has a BS from Tufts and an MS from Northeastern. He is an accountant who would rather be chasing or giving out counties on 20 meters. You can reach Tom at 54 Hermaine Ave, Dedham, MA 02026 or via **tom.hart@verizon.net**.

A 2 Meter and 70 CM Portable Tape Measure Beam

Work the OSCAR ham satellites or go transmitter hunting with this inexpensive portable dual band handheld tape measure Yagi.

John Portune, W6NBC

Many hams have made 2 meter tape measure beams for transmitter hunting. In this article we'll take that design two steps farther. First, we'll reduce cost and then we'll add OSCAR satellite Mode J capacity. A bonus is that a double boom design lets you fold the antenna up conveniently for transport or storage.

Keeping the Cost Down

When I first saw a tape measure 2 meter beam for transmitter hunting, I said: "What a great idea." So I bought a kit with tape measure material, PVC pipe, PVC X fittings and stainless steel hose clamps. It worked very well. Later I thought: "This would make a great ham club project if I could just get the cost down." Lightning struck — there's an easy approach that doesn't use expensive stainless steel hose clamps or PVC X fittings.

Simply drill 7/16 inch holes through a ½ inch inside diameter PVC boom and push the tape measure elements through. (See Figure 1.) It works just as well and is much less expensive. The long elements do need a short length of 3/8 inch wood or fiberglass dowel and a couple of tie wraps to stiffen them in the wind. The dowels push through the same holes. Now the most expensive parts of the beam is the RG-58 coax and the connector. Since then, local club members have made many 2 meter tape measure beams for transmitter hunting.

Working the OSCARs as Well

More recently I became interested in working Mode J OSCAR satellites — 2 meter FM phone toward the satellite, 70 cm FM phone from the satellite. What kind of antenna should I use? At a local club meeting I learned that there are several commercial handheld satellite beams on the market. The lecturer gave us a live demo of one during an actual satellite pass. But were there any good homebrew designs?

Then I remembered my handy little

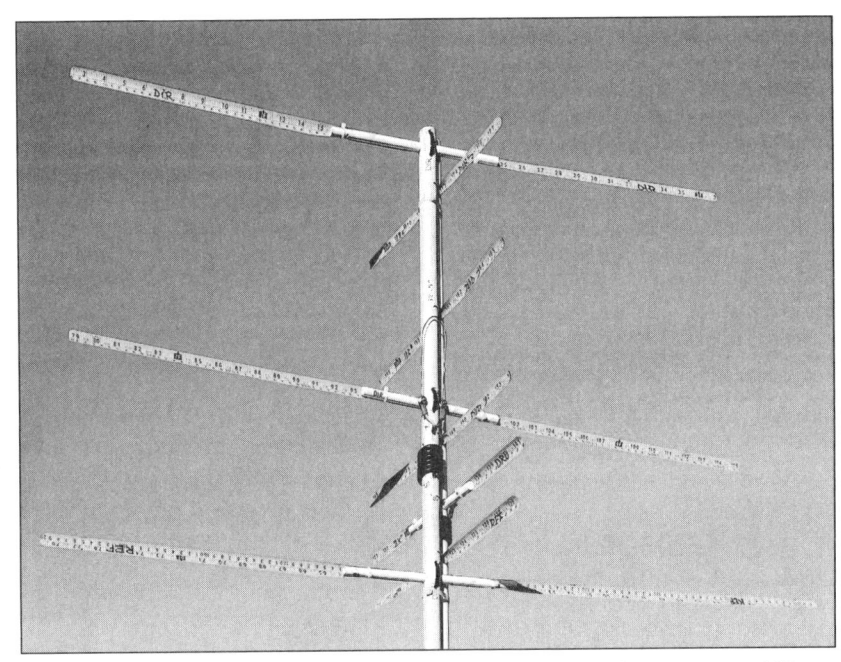

Figure 1 — Dual purpose, dual boom portable tape measure Yagi for 2 meters and 70 cm.

2 meter tape measure transmitter hunting beam. Why not just add a 70 cm beam to it? The one shown is the result. I did not make any changes to the existing three element 2 meter transmitter hunting beam. For though it is a little shorter than most commercial satellite beams, and is optimized more for front to back ratio, it works just fine for the "birds" and of course still for transmitter hunting. I did, however, optimize the new 70 cm beam for forward gain. Figure 2 shows *EZNEC* elevation plots for both, with the beams turned vertically.

As just a bit of simple theory, the 70 cm beam needs to be rotated axially 90° from the 2 meter beam. This makes the two invisible to each other. I've seen designs in which they are in the same plane. I tried this and found it unsatisfactory. The reason, I believe, is that a 2 meter Yagi has a third harmonic resonance near 70 cm. The directivity of the 2 meter Yagi on UHF is poor. If you mount the two beams in the same plane, the patterns clash. I modeled this with *EZNEC* and saw that it is a poor idea.[1]

Folding for Storage

A particularly handy feature of this design is the dual boom. I had at first thought to just drill extra holes in the existing single boom for the 70 cm Yagi. In terms of performance, that would have been fine. But on the 2 meter only version I had been folding the element ends back under tie wraps for transport or storage (see Figure 3). Had I added a 90° fixed position UHF beam, the whole structure would have become cumbersome. Two rotatable booms is a better approach. By holding the booms together with two tie wraps one can easily rotate the beams into

[1]Several versions of *EZNEC* antenna modeling software are available from developer Roy Lewallen, W7EL, at **www.eznec.com**.

Antenna System Projects   6-19

the same plane for storage or transport. Most of the figures in the article show them in that mode. The folded-up dual-band beam is now no bigger than before.

Compatible Element Spacing

One of the slightly tricky parts was to design a 70 cm beam with elements that fit well between the existing elements of the 2 meter beam. Fortunately, the spacing of Yagi elements isn't critical. Experience has shown me that one can select almost any spacing, within limits, and by simply then adjusting element length, you get pretty much the same performance. Figure 4 gives placements and sizes. The elements are made from ordinary small-width tape measure material.

Baluns and Feed Point

The impedance at the center of the 2 meter beam will be less than 50 Ω. By shortening the driven element and inserting an inductive hairpin in shunt with the feed point, the impedance will be raised to 50 Ω resistive. Make an inductive hairpin as shown in Figure 5 from #12 AWG solid copper wire, 8½ inches in length. It yields reasonable SWR and is easily adjusted. The 70 cm Yagi's driven element naturally matches 50 Ω with-

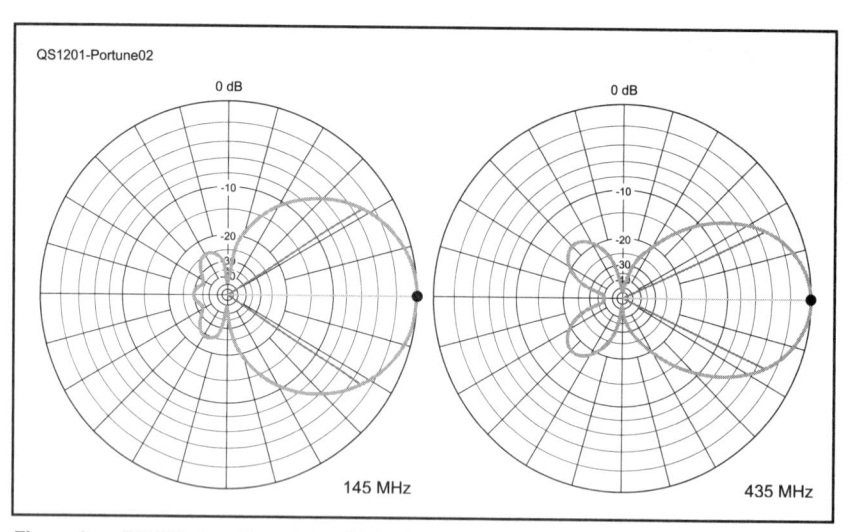

Figure 2 — *EZNEC* elevation plots with both Yagis vertically polarized. As shown, the 145 MHz gain is 7.5 dBi, the front-to-back ratio is 29 dB. For 435 MHz, the gain is 10.4 dBi, the front-to-back ratio is 38 dB.

Figure 3 — Booms rotated 90° and 2 meter elements folded back. Note tie wraps to hold ends.

> ### Hamspeak
>
> **Diplexer** — Passive device that accepts energy from two sources on different frequencies and combines them into signals on a single port. Alternately, it can accept signals on two frequencies combined into a single stream and separate them into signals on two ports based on their frequency.
>
> **EZNEC** — Antenna modeling software that provides a user friendly interface to the powerful *Numerical Electromagnetic Code* (NEC) calculating engine. Several versions of *EZNEC* antenna modeling software are available from developer Roy Lewallen, W7EL, at **www.eznec.com**.
>
> **OSCAR** — Orbiting Satellite Carrying Amateur Radio. Name given to a set of Amateur Radio satellites. The number following the name indicates the deployment sequence.
>
> **OSCAR Mode J** — Amateur satellite mode referring to a single channel FM repeating satellite that has a 2 meter uplink and a 70 cm downlink.
>
> **Yagi** — Multielement directive antenna array in which one or more elements are driven by connection to a transmission line and the others are parasitically coupled. Yagis are generally characterized by high gain for their size accompanied by narrow operating frequency range.

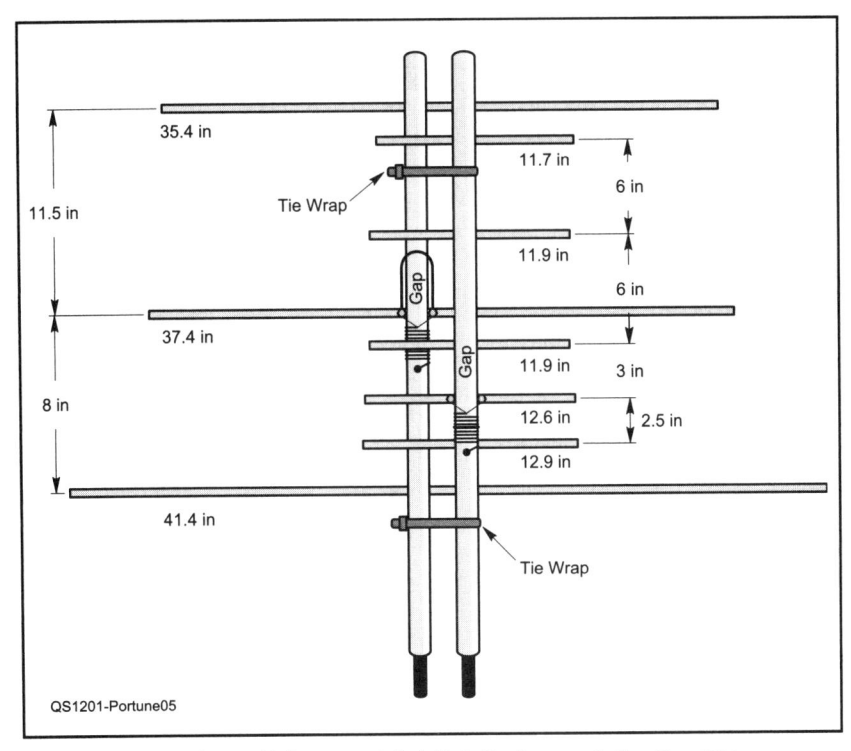

Figure 4 — Dimensions with booms rotated. Note the boom rotation tie wraps.

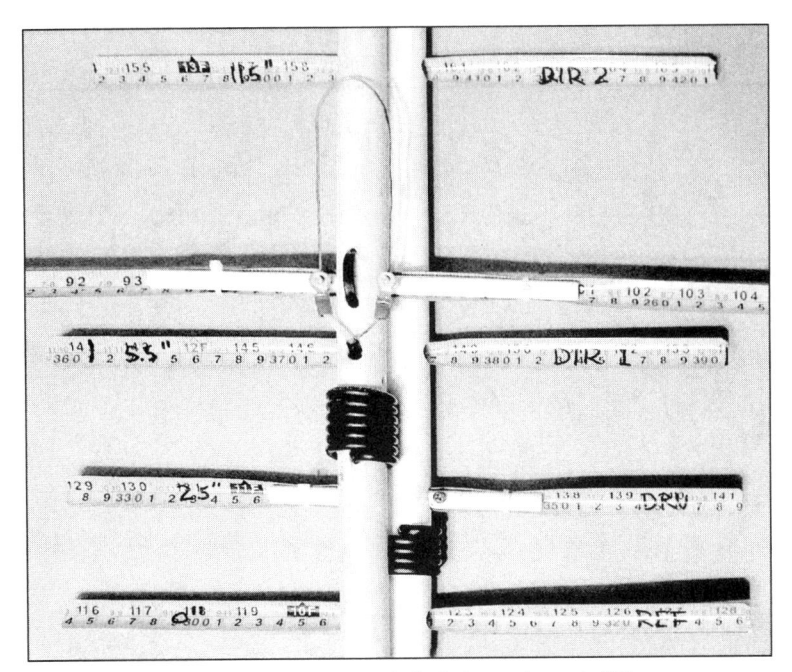

Figure 5 — Feed-point detail showing choke baluns, element stiffeners and hairpin match on the 2 meter beam. The booms have been rotated into storage mode.

out any added tuning elements.

As with most coax fed antennas, baluns on both beams is highly recommended. Shown are choke baluns made by wrapping the RG-58 feed coax around the boom several times. Secure the ends through holes in the boom. For 2 meters, six turns is adequate; for 70 cm, use four turns. Run the coax down the inside of the boom and out the end. Use two 6-32 screws and four nuts to connect to each driven element. Leave a ¼ inch gap in the middle. Scrape the paint off the tape measure material and tin the area with solder to provide a good connection. Separate out a short length of the coax's inner and outer conductors and crimp on ring terminals.

Using the Yagi with a Dual-band Radio

I normally make my OSCAR contacts using two separate handheld transceivers, one for each band. You may, of course, use a dual band handheld, but in that case you will need a diplexer to connect the two antenna feed lines to the radio's antenna port. There are also several satisfactory homebrew diplexer designs on the Internet.

I claim no special magic for this antenna design. The Yagis are classical designs and the tape measure method well known. The charm is the mechanical arrangement. By eliminating PVC X fittings and stainless hose clamps, it is inexpensive, and by also using two booms it folds up neatly. I now can seamlessly combine two of my favorite ham radio disciplines, transmitter hunting and ham satellites in one handy portable antenna.

Photos by the author.

ARRL Member John Portune, W6NBC, received a BSc in physics from Oregon State University in 1960, his FCC Commercial General Radiotelephone license in 1961 and his Advanced class amateur license in 1965. He spent five years in England as G5AJH and upgraded to Amateur Extra class in 1985. John retired, first as a broadcast television engineer and technical instructor at KNBC in Burbank and then from Sony Electronics in San Jose, California.

*John is active on many bands and modes, predominantly from his HF equipped RV mobile station. He has written various articles in ham radio and popular electronics magazines and remains active as a VE team leader, ham license teacher and website designer. You can reach John at 1095 W McCoy Ln #99, Santa Maria, CA 93455, or at jportune@aol.com or via his website at **w6nbc.com**.*

Antenna System Projects   6-21

An EZ-Lindenblad Antenna for 2 Meters

This easy to build antenna works well for satellite or terrestrial communication, horizontal or vertically polarized.

Anthony Monteiro, AA2TX

Lindenblad is the name of a type of antenna that is circularly polarized yet has an omnidirectional radiation pattern. With most of its gain at low elevation angles, it is ideal for accessing low earth orbit (LEO) Amateur Radio satellites. Because it is omnidirectional, it does not need to be pointed at a satellite so it eliminates the complexity of an azimuth/elevation rotator system. This makes the Lindenblad especially useful for portable or temporary satellite operations. It is also a good general purpose antenna for a home station because its circular polarization is compatible with the linearly polarized antennas used for FM/repeater and SSB or CW operation.

This type of antenna was devised by Nils Lindenblad of the Radio Corporation of America (RCA) around 1940.[1] At that time, he was working on antennas for the then nascent television broadcasting (TV) industry. His idea was to employ four dipoles spaced equally around a λ/3 diameter circle with each dipole canted 30° from the horizontal. The dipoles are all fed in phase and are fed equal power. The spacing and tilt angles of the dipoles create the desired antenna pattern when the signals are all combined. Unfortunately, the start of World War II halted Lindenblad's TV antenna work.

After the war, George Brown and Oakley Woodward, also of RCA, were tasked with finding ways to reduce fading on ground-to-air radio links at airports.[2] These links used linearly polarized antennas. The maneuverings of the airplanes often caused large signal dropouts if the antennas became cross-polarized. Brown and Woodward realized that using a circularly polarized antenna at the airport could reduce or eliminate this fading so they decided to try Lindenblad's TV antenna concept.

Brown and Woodward designed their antenna using metal tubing for the dipole elements. Each dipole element is attached to a section of shorted open-wire-line, also made from tubing, which serves as a balun transformer. A coaxial cable runs through one side of each open-wire-line to feed each dipole. The four coaxial feed cables meet at a center hub section where they are connected in parallel to provide a four-way, in-phase power-splitting function. This cable junction is connected to another section of coaxial cable that serves as an impedance matching section to get a good match to 50 Ω. While the Brown and Woodward design is clever and worked well, it would be quite difficult for the average ham (including this author!) to duplicate.

The major cause of the difficulty in designing and constructing Lindenblad antennas is the need for the four-way, in-phase, power splitting function. Since we generally want to use 50 Ω coaxial cable to feed the antenna, we have to somehow provide an impedance match from the 50 Ω unbalanced coax to the four 75 Ω balanced dipole loads.

Previous designs have used combinations of folded dipoles, open-wire lines, twin-lead feeds, balun transformers and special impedance matching cables in order to try to get a good match to 50 Ω. These in turn increase the complexity and difficulty of the construction.

[1]Notes appear on page 6-25.

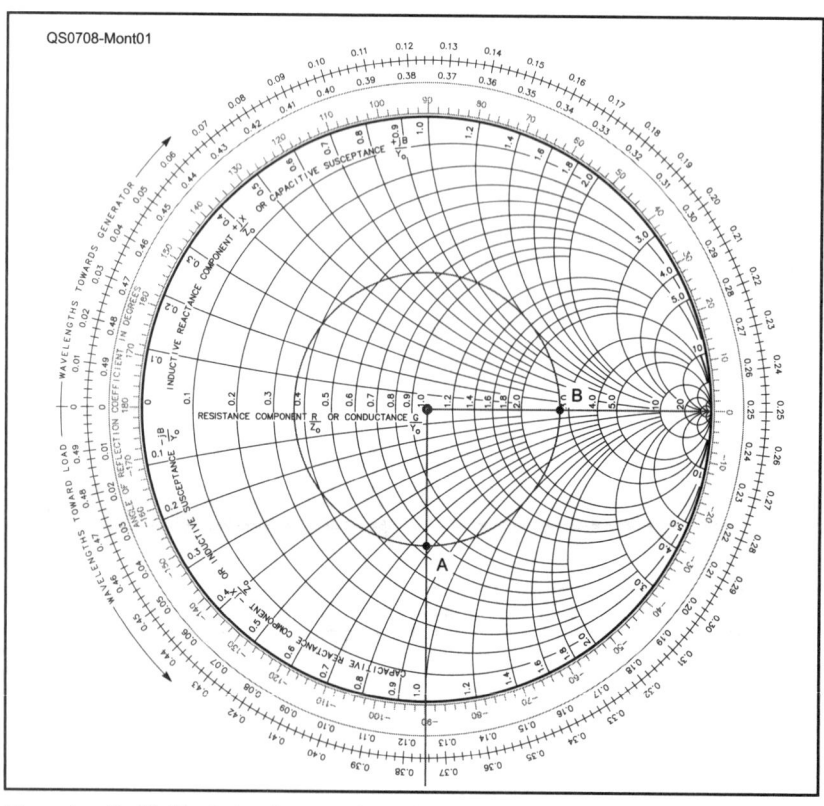

Figure 1 — Smith Chart showing transformation along a 75 Ω transmission line to yield 200 Ω.

Table 1
Required Materials

Quantity one, unless noted.
Aluminum tubing, 17 gauge, 6 foot length of ¾" outer diameter, quantity 3. Available from Texas Towers, **www.texastowers.com**.
Aluminum angle stock, 8" length of 2" × 2" × 1/16".
Aluminum angle stock, 2" length of 2" × 2" × 1/16" for mounting connector.
Screws, #8 × ½" aluminum sheet metal, quantity 12.
Screws, #8 × ½" aluminum sheet metal or 3/16" aluminum rivets, quantity 12.
PVC insert T-connector, ½" × ½" × ½" grey for irrigation polyethylene tubing. LASCO Fittings, Inc. Part# 1401-005 or equivalent. Available from most plumbing supply and major hardware stores, quantity 4.
Plastic end caps (optional), black ¾", quantity 8.
N-connector for RG-8 cable, single-hole, chassis-mount, female.
Cable ferrite, Fair-Rite part # 2643540002, quantity 4 (Mouser Electronics #623-2643540002).
RG-59 polyethylene foam coax with stranded center conductor, 10' length.
Copper braid, 4" long piece.
Ring terminal, uninsulated 22-18 gauge for 8-10 stud, quantity 4.
Ring terminal, uninsulated 12-10 gauge for 8-10 stud, quantity 4.
Heat shrink tubing for ¼" cable, wire ties, electrical tape, as needed.
Ox-Gard OX-100 grease for aluminum electrical connections.

The EZ-Lindenblad

An antidote to these difficulties is the EZ-Lindenblad. The key concept of the EZ approach is to eliminate anything that is electrically or mechanically difficult, leaving only things that are *easy*. This leads to the idea of just feeding the four dipoles with coax cable and soldering the cables to a connector with no impedance matching devices at all. This would certainly be *easy* but we also want the antenna to work! Without the extra impedance matching devices, how is it possible to get a good match to 50 Ω?

If we could get each of the four coax feed cables to look like 200 Ω at the connector, then the four in parallel would provide a perfect match to 50 Ω. We could do this if we used quarter-wavelength sections of 122 Ω coax to convert each 75 Ω dipole load to 200 Ω. Unfortunately, there is no such coax that is readily available.

But we can accomplish the same thing with ordinary 75 Ω, RG-59 TV type coax if we run the cable with an intentional impedance mismatch. By forcing the standing wave ratio (SWR) on the cable to be equal to 200/75, or about 2.7:1, we can make each cable look exactly like 200 Ω at the connector as long as we make them the right length. It is easy to make the SWR equal 2.7:1 by just making the dipoles a little too short for resonance. An *EZNEC* antenna model can be used to determine the exact dipole dimensions.[3]

Figure 2 — Close-up of dipole electrical connections.

Figure 3 — View of cross booms mounted to mast.

The conversion from the balanced dipole load to the unbalanced coax cable can be painlessly accomplished by threading each cable through an inexpensive and readily available ferrite sleeve making essentially a choke balun. The only remaining issue is the required length of the feed cables, and this can be easily determined using a *Smith Chart*.

Smith Chart

The Smith Chart was invented by Phillip Smith of The Bell Telephone Laboratories in 1939.[4] As a high school student, Phillip had been a ham radio operator and used the call sign 1ANB. After graduating from Tufts College, he went to work for Bell Labs in the radio research department. As part of his job, he needed to make many impedance calculations that in those days required doing many complex computations by hand. Phillip realized he could create a chart that would allow the solution to be plotted on a graph making his job a lot easier. Phillip Smith's chart was soon adopted by other radio engineers and quickly became a standard engineering tool that is still in use today. Technical information and free downloadable Smith Charts are available via links on the ARRL Web site.[5] Please see the Smith Chart in Figure 1.

With a Smith Chart, we can easily determine the required length of 75 Ω coax to provide a 200 Ω load. The Smith Chart shown is normalized to 1 Ω in the center so we must multiply all impedance values by our coax impedance of 75 Ω. An *EZNEC* antenna model was used to simulate cutting the dipole lengths until the SWR on the line reached 2.7:1. The model showed that the dipole load impedance would then be 49 –j55 Ω and this is plotted at point A on the chart. The desired 200 Ω impedance at the connector is plotted at point B on the chart. A constant 2.7:1 SWR curve is drawn between the two impedance points. The length of cable needed is read clockwise along the scale labeled, WAVELENGTHS TOWARD GENERATOR from the lines drawn through points A to B. From the chart, the length of the line needed is 0.374 λ.

The EZ-Lindenblad was designed for a center frequency of 145.9 MHz to optimize its performance in the satellite sub band. At 145.9 MHz, a wavelength is about 81 inches and since the coax used has a velocity factor of 0.78, we need to make the feed cables 81 × 0.374 × 0.78 = 23.6 inches long.

There are several computer programs available today that can also be used to do the Smith Chart calculations. These include *TLW*, provided with *The ARRL Antenna Book* and *MicroSmith* formerly offered by the ARRL, which was used as a cross check.[6]

Construction

This antenna was designed to be rugged and reliable yet easy to build using only hand

Antenna System Projects   6-23

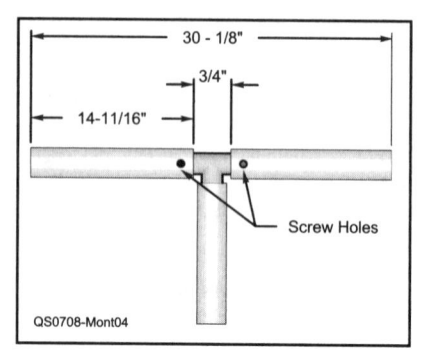

Figure 4 — Dipole dimensions.

Table 2
Aluminum Tubing Lengths

Description	Quantity	Length
Dipole rods	8	14¹¹⁄₁₆"
Cross booms	2	23"

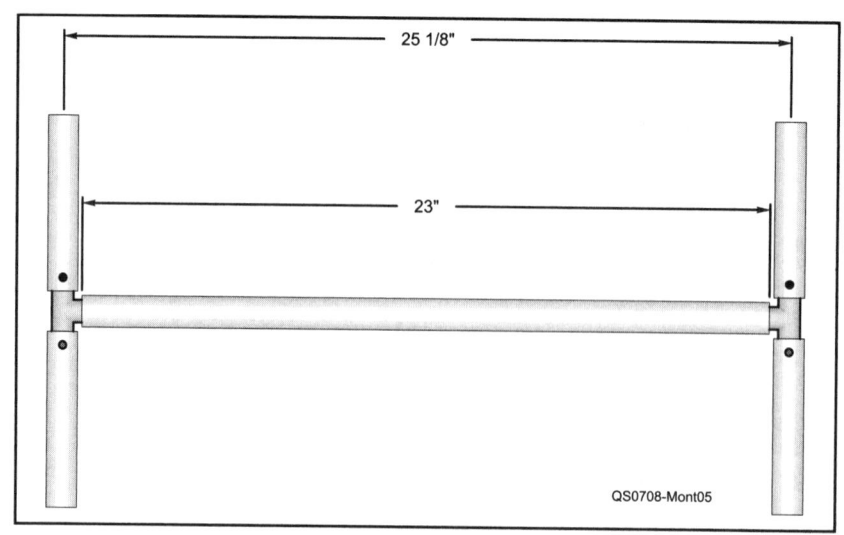

Figure 5 — Dimension of dipole spacing on cross-boom.

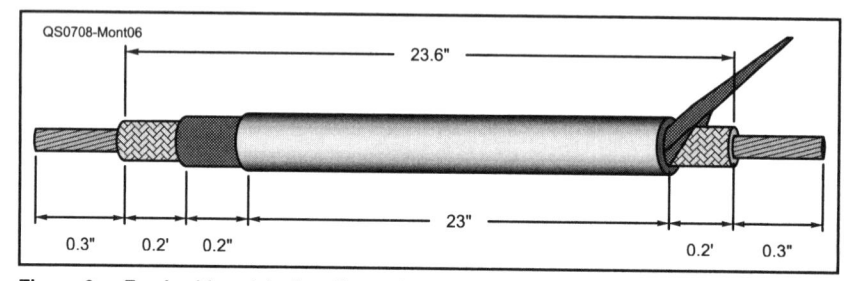

Figure 6 — Feed cables stripping dimensions.

tools with all of the parts readily available as well. Please see the parts list in Table 1. Although not critical, the construction will be easier if the specified 17 gauge aluminum tubing is used since the inner wall of the tubes will be just slightly smaller than the outer wall of the PVC insert Ts used to connect them. If heavier gauge tubing is used, it will be necessary to file down the PVC insert Ts to make them fit inside the aluminum tubes.

Start by making a mounting bracket to mount the N-connector and the cross booms. Cut a ⅝ inch hole in one side of the short piece of angle stock and rivet or screw it to the bottom of the long piece of angle stock. The completed bracket with the connector and cables attached can be seen in Figure 3.

Next, cut the aluminum tubing to make the cross booms and dipole rods as shown in Table 2. Drill holes for the machine screws at each end of the cross booms but do not insert the screws yet. Attach the cross booms to the long section of angle stock with rivets or screws. One cross boom will mount just above the other as can be seen in Figure 3. The cross booms should be perpendicular to the mounting bracket so that they will be horizontal when the antenna is mounted to its mast. Make sure that the centers of the cross booms are aligned with each other so that the ends of the cross booms are all 11½ inches from the center cross.

Make the dipoles by inserting a PVC insert T into two dipole rods. It should be possible to gently tap in the rods with a hammer but it may be necessary to file down the insert T a little if the fit is too tight. Applying a little PVC cement to the insert T will soften the plastic and make it easier to insert into the aluminum tubing if the fit is too tight. The overall dipole length dimension is critical so take care to get this correct as shown in Figure 4.

Drill holes for machine screws in each dipole rod but do not insert the screws yet. The screws will be used to make the electrical connections to the dipoles at the center. The screw holes should be about ⅜ inch from the end of the tubing.

The dipole assemblies are attached by gently tapping the PVC insert-T into the end of each cross-boom with a hammer. The dimensions are shown in Figure 5.

Next, temporarily attach the mounting bracket to a support so that each of the cross booms is perfectly horizontal. Measure this with a protractor. Now, using the protractor, rotate the dipole assemblies to a 30° angle with the right-hand side of the nearest dipole tilting up when you are looking toward the center of the antenna. Drill a small hole through the existing cross-boom holes into the PVC insert-Ts and then use the machine screws to fasten the dipole assemblies into place. For a nice finishing touch, the dipole ends can be fitted with ¾ inch black plastic end caps.

Next, make the four feed cables by cutting and stripping the RG-59 coax as shown in Figure 6. On the dipole connection side, unwrap the braid and form a wire lead. Apply the smaller ring terminal to the center conductor and use the larger ring terminal for the braid. At the other end of the cable, do not unwrap the braid but strip off the outer insulation. Slip a 1 inch piece of shrink wrap over the coax and apply to the dipole side. Next,

Figure 7 — The EZ-Lindenblad as a portable or Field Day antenna. A 70 cm antenna is on the top.

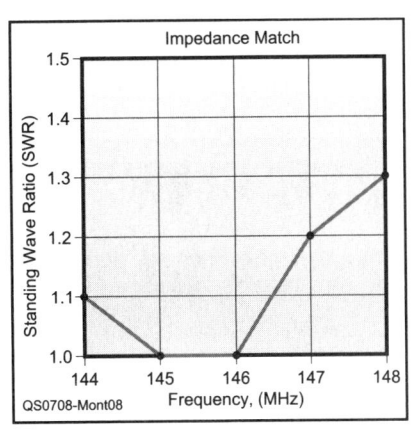

Figure 8 — Lindenblad antenna SWR at 50 Ω.

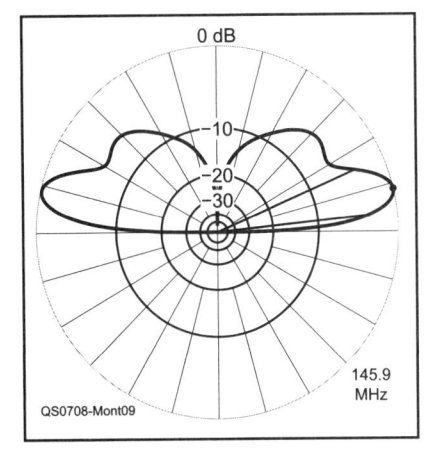

Figure 9 — *EZNEC* radiation pattern of Lindenblad antenna.

slip a cable ferrite over the cable and push all the way to the dipole end as far as it will go (ie, up to the heat-shrink tubing.) The fit will be snug and you may need to put a little grease on the cable jacket to get it started.

Prepare each dipole for its feed cable by first cleaning the area around the screw holes with steel wool and then applying Ox-Gard grease. This is to ensure a good electrical connection. The coax center conductor goes to the up side of the dipole and the braid goes to the down side. To make a connection, put a machine screw through the ring terminal and gently screw into the dipole tubing. Do not overtighten the screws or you will strip the tubing.

Apply Ox-Gard or other corrosion resistant electrical grease around the hole for the N connector. Take the 4 inch piece of braid and put the end of it through the hole for the N connector. This provides the ground connection. Secure the N connector in the mounting hole to clamp the braid. Use a wire tie or tape to hold the four feed cables together at the connector ends. Make sure to align the cables so that all the ground braids are together and the center conductors all extend out the same amount. Do not twist the center conductors together. Carefully push the four cable center conductors into the center terminal of the N connector and solder them in place. Wrap the exposed center conductors of the cables and the connector with electrical tape.

Take the piece of braid that is clamped to the N connector and wrap it around the four exposed ground braids of the coax cables. Solder them all together. This will take a fair amount of heat but be careful not to melt the insulation. After this cools, apply electrical tape over all the exposed braid and fix with wire ties. The cables should be secured to the cross booms with wire ties.

The mounting bracket provides a way to attach the antenna to a mast using whatever clamping mechanism is convenient. For a permanent mount, drill holes in the bracket to accept a pair of U bolts. The author's antenna was intended for portable operation and the bracket was drilled to accept two #8 stainless steel screws. These screws pass through a portable mast and the antenna is secured with stainless steel thumbscrews. This allows the antenna to be set up or taken down in less than a minute. The completed portable antenna as used for Field Day 2006 is shown in Figure 7. The little antenna at the top is for 70 cm.

Standing Wave Ratio

The antenna impedance match to 50 Ω was tested using an MFJ-259B SWR meter, which has a digital readout of standing wave ratio (SWR) and frequency. The frequency accuracy was verified using an external frequency counter and the 1.0:1 calibration was checked with a Narda precision 50 Ω load.

The antenna was connected to the SWR meter with a 6 foot coax jumper made of Belden 9913F7, which has very low loss. The SWR was measured at 1 MHz intervals over the 144-148 MHz range. As can be seen in the chart of Figure 8, the antenna provides an excellent match over the entire 2 meter band.

Power Handling Capability

This antenna was designed to safely handle any of the currently available VHF transceivers. The power handling capability was tested by applying a 200 W CW signal, key down for 9½ minutes. Immediately after the test, the ferrites and cables were checked and there was no noticeable temperature rise.

Radiation Pattern

The antenna radiation pattern predicted by the *EZNEC* model is shown in Figure 9. This is the elevation plot with the antenna mounted at six feet above ground although it can be mounted higher if desired for better coverage to the horizon. As shown in the plot, the pattern favors the lower elevation angles. The –3 dB points are at 5° and 25° with the maximum gain of 4.8 dBic (with respect to an isotropic circularly polarized antenna) at around 13°. This is an excellent pattern for accessing LEO satellites. Most of the satellite pass elevations will be in this range and it is also the elevation at which the satellite provides the best chance for DX contacts.

The antenna radiation is right-hand circularly polarized, which will work with virtually any LEO satellite that uses the 2 meter band. The circularity was checked by measuring the difference between the horizontal and vertical radiation components. This was done using a linearly polarized sense antenna mounted 100 feet away feeding into an FT-817 radio with the AGC switched off. The radio audio was connected to an ac voltmeter with a dB scale. The test showed a difference of less than 3 dB, which is very good for an omnidirectional antenna. A reference horizontally polarized antenna measured nearly a 30 dB difference.

On the Air

The EZ-Lindenblad antenna has been used for SSB, FM and packet operation on the AO-07, FO-29, SO-50, AO-51, VO-52, NO-44, NO-60 and NO-61 satellites. The portable setup, as seen in Figure 5, was used for the satellite station at the North Shore Radio Association, NS1RA, 2006 Field Day effort. Field Day is an excellent test of any antenna as it is probably the busiest 48 hours of the year on the satellites and the EZ-Lindenblad performed well.

An earlier version of this antenna was published in the *AMSAT Space Symposium Proceedings* of October 2006.

Notes

[1] G. Brown and O. Woodward Jr. *Circularly Polarized Omnidirectional Antenna*, RCA Review, Vol 8, no. 2, Jun 1947, pp 259-269.
[2] See Note 1.
[3] *EZNEC+ V4* antenna analysis software by Roy W. Lewallen, W7EL, available from **www.eznec.com**.
[4] F. Polkinghorn, "Phillip H. Smith, Electrical Engineer, an Oral History," IEEE History Center, Rutgers University, New Brunswick, NJ. Available at **www.ieee.org**.
[5] Smith Chart information is available at **www.arrl.org/tis/info/chart.html**.
[6] R. D. Straw, Editor, *The ARRL Antenna Book*, 21st Edition. Available from your ARRL dealer or the ARRL Bookstore, ARRL order no. 9876. Telephone 860-594-0355, or toll-free in the US 888-277-5289; **www.arrl.org/shop/**; **pubsales@arrl.org**.

Index

A
Amateur Radio on the International Space Station 5-5
   School Contacts 5-15
Amplifiers, RF Power..3-15
Antennas 6-2
   A 2 Meter and 70 CM Portable Tape Measure
      Beam (project)...6-19
   A Simple Fixed Antenna for VHF/UHF Satellite
      Work (project) 6-2
   An EZ-Lindenblad Antenna for 2 Meters (project) 6-22
   Arrow Portable Yagi 3-2
   Circularly Polarized Yagi Antennas for Satellite
      Communications (project)...6-13
   Diplexer 3-2
   Dual Band Handy Yagi (project) 6-17
   Eggbeater 3-5
   Elk Portable Log Periodic 3-2
   Gain and Size 3-9
   Lindenblad 3-6
   Omnidirectional..3-4
   Polarization 3-7
   Quadrifilar 3-7
   Rotators 3-10
   Turnstile 3-5
Azimuth...2-6
Azimuth/Elevation Antenna Rotator...3-10

D
Diplexer 3-2
Doppler Effect...2-9

E
Education..5-1
   Amateur Radio on the International Space Station 5-5
   Promoting Amateur Radio 5-1
   Research Satellites..5-4
   The FUNcubes 5-3
Eggbeater Antenna...3-5
Elevation...2-6

F
Footprints 2-5
FUNcubes 5-3

G
Grid Squares 2-2

H
History, Satellites 1-1

I
International Space Station 5-5
    HamTV 5-14
    Packet 5-12
    QSLs 5-14
    School Contacts 5-15
    SSTV 5-13
    Voice Contacts 5-12

L
Lindenblad Antenna 3-7
Linear Transponders 4-9
Logbook of The World 4-6

M
Memories, Programming 4-3

O
Omnidirectional Antennas 3-4
Orbital Elements 2-14
Orbits 2-3

P
Polarization, Antenna 3-7
Preamplifiers 3-12
Projects
    A 2 Meter and 70 CM Portable Tape Measure Beam 6-19
    A Simple Fixed Antenna for VHF/UHF Satellite Work 6-2
    An Affordable Az-El Positioner for Small Antennas 6-6
    An EZ-Lindenblad Antenna for 2 Meters 6-22
    Circularly Polarized Yagi Antennas for Satellite
        Communications 6-13
    Dual Band Handy 6-17

Q
Quadrifilar Antenna 3-7

R

Rotators 3-10
   An Affordable Az-El Positioner for
      Small Antennas (project) 6-6
   Azimuth/Elevation 3-10

S

Satellites
   Doppler Effect 2-9
   FM Repeater 4-2
      Making Contacts 4-5
      Programming Memories 4-3
   Footprints 2-5
   History 1-1
   Linear Transponder 4-9
   Orbits 2-3
   Tracking 2-1
      Azimuth and Elevation 2-6
      Orbital Elements 2-14
      Software 2-12
      Websites 2-18
Software 2-12
Stations
   Receive Preamplifiers 3-12
   RF Power Amplifiers 3-15
   Transceivers 3-13

T

Tracking 2-1
   Azimuth and Elevation 2-6
   Orbital Elements 2-14
   Software 2-12
   Websites 2-18
Transceivers 3-13
   Programming Memories 4-3
Turnstile Antenna 3-5